T0313895

Analytic Theory
of Global Bifurcation

PRINCETON SERIES IN APPLIED MATHEMATICS

TITLES IN THE SERIES

Chaotic Transitions in Deterministic and Stochastic Dynamical Systems: Applications of Melnikov Processes in Engineering, Physics and Neuroscience by Emil Simiu

Selfsimilar Processes by Paul Embrechts and Makoto Maejima

Self-Regularity: A new Paradigm for Primal-Dual Interior Points Algorithms by Jiming Peng, Cornelis Roos and Tamás Terlaky

Analytic Theory of Global Bifurcation: An Introduction by Boris Buffoni and John Toland

The Princeton Series in Applied Mathematics publishes high quality advanced texts and monographs in all areas of applied mathematics. Books include those of a theoretical and general nature as well as those dealing with the mathematics of specific applications areas and real-world situations.

Analytic Theory
of Global Bifurcation

An Introduction

Boris Buffoni and John Toland

PRINCETON UNIVERSITY PRESS

PRINCETON AND OXFORD

Published by Princeton University Press,
41 William Street, Princeton, New Jersey 08540

In the United Kingdom: Princeton University Press,
3 Market Place, Woodstock, Oxfordshire OX20 1SY

ISBN 0-691-11298-3 (cloth : alk. paper)

British Library Cataloging-in-Publication Data has been applied for

This book has been composed in Times
by Stephen I. Pargeter, Banbury, Oxfordshire, UK

Printed on acid-free paper. ⊗

www.pupress.princeton.edu

Printed in the United States of America

10   9   8   7   6   5   4   3   2   1

Contents

Preface

In the study of homogeneous linear problems of the form $Lu = 0$, the amplitude (or size) of a solution is unimportant as non-zero solutions are never isolated and the existence of one non-zero solution implies the existence of others with all possible amplitudes. This is not the case for nonlinear equations of the form $N(u) = 0$ and the observation that a solution is unique in a neighbourhood of itself has no ramifications globally for the set of all solutions of the equation.

Since nonlinear equations are often derived without restrictions on the amplitudes of likely solutions, with the intention of describing large amplitude phenomena, it is essential to have mathematical methods which discover the existence of solutions without regard to their size. These notes are concerned with analytical aspects of this general question.

Global bifurcation theory deals with the existence *in-the-large* of connected sets of solutions to nonlinear equations of the form

$$F(\lambda, x) = 0, \quad \lambda \in \mathbb{R}, \quad x \in X \setminus \{0\},$$

where $F : \mathbb{R} \times X \to Y$, X, Y are Banach spaces and $F(\lambda, 0) \equiv 0$. It is well known that P. H. Rabinowitz's now-classical topological theory of global bifurcation leads to the existence of sets of solutions which are connected, but not in general path-connected even when the operators involved are infinitely differentiable. On the other hand, E. N. Dancer pointed out that connectedness in the topological theory can be replaced with path-connectedness if the operators are real-analytic. The two approaches from the early 1970s are completely different.

Quite recently, in a collaboration with Dancer on a problem from hydrodynamic wave theory, we encountered the following situation. If the existence globally of a path of solutions of a certain real-analytic equation could be established, it was clear from earlier work of P. I. Plotnikov that along the path the Morse index must increase without bound; this in turn would lead to infinitely many secondary bifurcation points and ultimately to the existence of sub-harmonic bifurcations, which was our goal.

Here, as in many similar problems, the existence globally of paths (as opposed to connected sets) of solutions is the key. Our purpose therefore is to give a self-contained account of such a theory, and in particular to focus on the existence of global paths of solutions as a consequence of bifurcation from a simple eigenvalue in the real-analytic case.

There follows an expanded version of the notes for a course of postgraduate lectures on real-analytic global bifurcation theory and its applications which we gave as part of *'un cour de 3ème cycle à l'Ecole Polytechnique Fédérale de Lausanne'* during the winter term of the academic year 1999-2000. We have benefited greatly from the interest of those who attended the lectures, asked questions, read drafts and offered advice: in particular we thank A. André, D. Crispin, A. Pichler-Tennenberg, F. Gebran, S. Rey and C. A. Stuart. We must also thank E. N. Dancer for his encouragement and comments on the manuscript.

We assume a knowledge of undergraduate linear functional analysis and of calculus notation in finite dimensions. We also assume a knowledge of a first course in functions of one complex variable and some elementary linear algebra. However calculus in Banach spaces is treated with complete proofs, from the definition of a Fréchet derivative to the inverse and implicit function theorems. We include an account of how infinite-dimensional problems can be reduced locally to equivalent ones in finite dimensions and of how that leads to standard results from local bifurcation theory. This theory assumes only that operators have a specified finite degree of differentiability.

Then we develop the theory of infinite-dimensional analytic operators over $\mathbb{R}$ or $\mathbb{C}$ and re-prove the inverse and implicit function theorems in that context. The elementary parts of the theory of finite-dimensional analytic varieties are also developed from first principles.

Finally, we study the global theory of *real-analytic, one-dimensional continua of solutions*. This has been the main goal, but the methods and results discussed here have much wider applicability to nonlinear operator equations. The application to the water-wave problem is considered in some detail.

A list of notation is included in the index.

Acknowledgements: Boris Buffoni held a grant from the Swiss National Science Foundation and John Toland was a Senior Fellow supported by the UK's EPSRC during the preparation of this manuscript.

Boris Buffoni, Lausanne
John Toland, Bath

31 March 2002

Chapter One

Introduction

Consider a system of k scalar equations in the form

$$F(\lambda, x) = 0 \in \mathbb{F}^k, \tag{1.1}$$

where $x \in \mathbb{F}^n$ represents the state of a system and $\lambda \in \mathbb{F}^m$ is a vector parameter which controls x. (Here $\mathbb{F}$ denotes the real or complex field.) A solution of (1.1) is a pair $(\lambda, x) \in \mathbb{F}^m \times \mathbb{F}^n$ and the goal is to say as much as possible qualitatively about the solution set.

Since (1.1) is a finite-dimensional nonlinear equation it might seem unnecessarily restrictive or even pointless to distinguish between the λ and x variables. Why not instead write $(\lambda, x) = Z \in \mathbb{F}^{m+n}$ and study the equation $F(Z) = 0$ where singularity theory is all that is needed? For example, when $F : \mathbb{C}^{m+n} \to \mathbb{C}^k$ is given by a power series expansion (that is, F is analytic), a solution Z_0 is called a bifurcation point if, in every neighbourhood of Z_0, the solutions of $F(Z) = 0$ do not form a smooth manifold. Locally the solutions form an analytic variety, a finite union of analytic manifolds of possibly different dimensions. So the qualitative theory of $F(Z) = 0$ in complex finite dimensions is reasonably complete.

However (i) in our applications λ is a parameter and the dependence on λ of the solution set is important; (ii) we are looking for a theory that gives the existence globally (i.e. not only in a neighbourhood of a point) of connected sets of solutions; (iii) we are particularly interested in the infinite-dimensional equation

$$F(\lambda, x) = 0 \tag{1.2}$$

when X and Y are real Banach spaces, $F : \mathbb{R} \times X \to Y$ is real-analytic and

$$F(\lambda, 0) \equiv 0.$$

Let

$$\mathcal{S}_\lambda = \{x \in X : F(\lambda, x) = 0\}.$$

The set $\mathcal{S}_\lambda$ normally depends on the choice of λ and usually varies continuously as λ varies. However, it sometimes happens that there is an abrupt change, a bifurcation, in the solution set, as λ passes through a particular point λ_0. For example, in Figure 1.1 the number of solutions changes from one to two as λ increases through λ_0. For a general treatment of bifurcation theory, see [19].

At this stage it is useful to see an infinite-dimensional example in which the global solution set can be found explicitly.

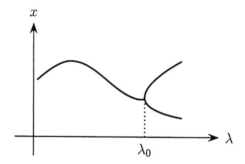

Figure 1.1 The set S_λ splits in two as λ passes through λ_0.

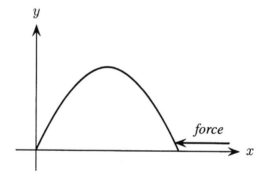

Figure 1.2 The rod bends under the action of a force.

1.1 EXAMPLE: BENDING AN ELASTIC ROD I

Consider an elastic rod of length $L > 0$ with one end fixed at the origin of the (x, y)-plane and with the other free to move on the x-axis under the influence of a force along the x-axis towards the origin. If we suppose that the length of the rod does not change (that it is incompressible) and if the force is big enough, then the rod will bend (see Figure 1.2).

We suppose that the rod always lies in the (x, y)-plane (there is no twisting out of the plane in the simple model which follows). To describe the rod's configuration let $(x(s), y(s))$ be the coordinates of a point at distance s (measured along the rod) from the end which is fixed at the origin. Since

$$x(s) = \int_0^s \cos\phi(t)dt \ \text{ and } \ y(s) = \int_0^s \sin\phi(t)dt,$$

the shape of the rod is given by the angle $\phi(s)$ between the tangent to the rod and the horizontal at the point $(x(s), y(s))$, $s \in [0, L]$.

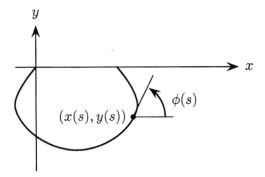

Figure 1.3 The angle between the tangent and the horizontal.

Let P denote the applied force. Then the Euler-Bernoulli theory [5, 6] of bending says that the curvature of the rod at a point is proportional to the moment created by the force. In other words,

$$-k\phi'(s) = Py(s),$$

where k, the constant of proportionality, is determined by the material properties of the rod, $Py(s)$ is the moment of the applied force and $-\phi'(s)$ is the curvature at the point $(x(s), y(s))$. It follows that if $P = 0$ then ϕ must be constant, and that constant must be 0 (mod 2π) since $y(L) = 0$. From now on we consider only the case $P > 0$. Since $y'(s) = \sin \phi(s)$ and $y(0) = y(L) = 0$ this gives

$$\phi''(s) + \lambda \sin \phi(s) = 0, \ s \in [0, L], \quad \phi'(0) = \phi'(L) = 0, \tag{1.3}$$

where $\lambda = P/k > 0$. If ϕ is a solution of (1.3), then so is $2k\pi + \phi$, for any $k \in \mathbb{Z}$. We therefore assume that $\phi(0) \in (-\pi, \pi)$. (If $\phi(0) = \pm\pi$ then ϕ is a constant.)

For all $\lambda > 0$, $(\lambda, \phi) = (\lambda, 0)$ is a solution of (1.3). This means that the mathematical model of bending admits a solution representing a *straight* rod, irrespective of how large the applied force might be. These solutions, $\phi = 0$, $\lambda > 0$ arbitrary, comprise the family of *trivial solutions*. To be realistic the model must also have solution corresponding to a bent rod (such as depicted in Figures 1.2 and 1.3). Note that any solution of (1.3) must satisfy the identity

$$\phi'(s)^2 + 4\lambda \sin^2(\tfrac{1}{2}\phi(s)) = 4\lambda \sin^2(\tfrac{1}{2}\phi_0), \quad s \in [0, L], \tag{1.4}$$

where $\phi_0 = \phi(0)$. This means that $(\phi(s), \phi'(s))$, $s \in [0, L]$, lies on a segment of the curve in (ϕ, ϕ')-phase space (see Figure 1.4) given implicitly by

$$\{(\phi, \phi') \in \mathbb{R}^2 : \phi'^2 + 4\lambda \sin^2 \tfrac{1}{2}\phi = 4\lambda \sin^2 \tfrac{1}{2}\phi_0\} \subset \mathbb{R}^2.$$

We therefore see that there is a solution joining $(-|\phi_0|, 0)$ to $(|\phi_0|, 0)$ in the half-space $\{(\phi, \phi') \in \mathbb{R}^2, \ \phi' \geq 0\}$ and one joining $(|\phi_0|, 0)$ to $(-|\phi_0|, 0)$ in the

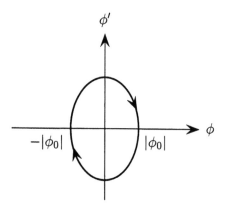

Figure 1.4 The direction of solutions in phase space.

half-space $\{(\phi, \phi') \in \mathbb{R}^2, \ \phi' \leq 0\}$, the corresponding value of L being given by the formula

$$L = \int_0^L \frac{|d\phi|}{|d\phi|/ds} = \int_{-|\phi_0|}^{|\phi_0|} \frac{d\phi}{\sqrt{4\lambda \sin^2 \frac{1}{2}\phi_0 - 4\lambda \sin^2 \frac{1}{2}\phi}}$$

$$= \frac{1}{\sqrt{\lambda}} \int_{-\pi/2}^{\pi/2} \frac{d\theta}{\sqrt{1 - \sin^2 \frac{1}{2}|\phi_0| \sin^2 \theta}},$$

where $\theta \in [-\pi/2, \pi/2]$ is given by $\sin(\phi/2) = \sin(|\phi_0|/2) \sin \theta$. In fact there are other solutions of (1.4) which in Figure 1.4 go around the curve $\frac{1}{2}K$ times for any positive integer K. For such solutions

$$L = \frac{K}{\sqrt{\lambda}} \int_{-\pi/2}^{\pi/2} \frac{d\theta}{\sqrt{1 - \sin^2 \frac{1}{2}|\phi_0| \sin^2 \theta}}.$$

This integral increases in $|\phi_0|$ and converges to $+\infty$ as $|\phi_0| \to \pi$.

Since L is the given length of the rod, this relation for each K is an implicit relation between $\phi_0 = \phi(0)$ and λ when (λ, ϕ) is a solution of (1.3). We can best describe the situation with the aid of a bifurcation diagram in which λ is the horizontal axis, ϕ_0 is the vertical axis, and L is fixed, see Figure 1.5.

The different curves correspond to different values of K, and it is easily checked that the Kth curve intersects the horizontal axis at $(K\pi/L)^2$.

It is fortunate but unusual that (1.3) can be reduced to (1.4) and that L can be calculated in terms of elliptic integrals. Because of this, solutions to (1.3) of all amplitudes can be found more-or-less explicitly. This is not the case for slightly more complicated problems and almost never for partial differential equations (PDEs).

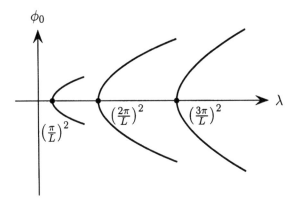

Figure 1.5 Bifurcation diagram.

General methods suitable for PDE applications are based on a study of (1.2). To put (1.3) in such a setting let

$$X = \{\phi \in C^2[0, L] : \phi'(0) = \phi'(L) = 0\},$$
$$Y = C[0, L] \text{ and } F(\lambda, \phi) = \phi'' + \lambda \sin \phi \in Y,$$

for all $(\lambda, \phi) \in \mathbb{R} \times X$. Then $F : \mathbb{R} \times X \to Y$ is smooth (Chapter 3) and real-analytic (Chapter 4). In Chapter 8 we show how equation (1.2) can be reduced locally to a finite-dimensional problem. This means that if (λ_0, x_0) satisfies (1.2) then there is a neighbourhood U of (λ_0, x_0) in $\mathbb{R} \times X$, a neighbourhood V of $(\lambda_0, 0) \in \mathbb{R} \times \mathbb{R}^N$ and an equation

$$f(\lambda, z) = 0 \in \mathbb{R}^M, \ (\lambda, z) \in \mathbb{R} \times \mathbb{R}^N, \ N, M \in \mathbb{N},$$

such that the solutions of the two equations are in one-to-one correspondence. The reduction to finite dimensions in §8.2 is called Lyapunov-Schmidt reduction and leads immediately to a local bifurcation theory based on the implicit function theorem. In particular, it yields a classical relation between a nonlinear problem and its linearization.

1.2 PRINCIPLE OF LINEARIZATION

Roughly speaking, the principle of linearization [39] derives from the feeling that when $F(\lambda, 0) = 0$ for all λ and solutions with $\|x\|$ small are sought, the non-linear problem $F(\lambda, x) = 0$ might as well be replaced with the linear equation $\partial_x F[(\lambda, 0)]x = 0$, where $\partial_x F[(\lambda, 0)]$ denotes the linearization of F with respect to x at $x = 0$. Since $\sin \phi = \phi + O(|\phi|^3)$ as $\phi \to 0$, the linearization of the elastic-rod problem at $(\lambda_0, 0)$ is

$$\phi''(s) + \lambda_0 \phi(s) = 0, \ s \in [0, L], \ \phi'(0) = \phi'(L) = 0, \ \lambda_0 > 0,$$

and this problem has non-trivial solutions if and only if

$$\lambda_0 = (K\pi/L)^2 \text{ with } \phi(s) = \cos(K\pi s/L), \ \ K \in \mathbb{N}.$$

The question is, can any inference be drawn from this about the nonlinear problem (1.3)? The answer is that in quite general situations (including equation (1.3) as a special case) λ_0 is a bifurcation point on the line of trivial solutions of (1.2) only if the linearized problem $d_x F[(\lambda_0, 0)]x = 0$ has a non-trivial solution. The fact that this is also sufficient for bifurcation from the line of trivial solutions for (1.3) (but not in general) is a consequence of the theory of bifurcation from a simple eigenvalue, see §8.4 and §8.5.

1.3 GLOBAL THEORY

It is clear from Figure 1.5 that there is more to the solution set of equation (1.3) than is predicted by local theory. Global features of the diagram are not a consequence of finite-dimensional reduction methods alone. We will see in Chapter 9 how local bifurcation theory, the implicit function theorem and some elementary results on real-analytic varieties can be used to piece together a global picture of the solution set of (1.2), without assumptions about the size of the solutions under consideration. Provided some general functional-analytic structure is present and F is real-analytic, the global continuum $\mathcal{C}$ of solutions which bifurcates from the trivial solutions at a simple eigenvalue contains a continuous curve $\mathfrak{R}$ with the following properties.

- $\mathfrak{R} = \{(\Lambda(s), \kappa(s)) : s \in [0, \infty)\} \subset \mathcal{C}$ is either unbounded or forms a closed loop in $\mathbb{R} \times X$.
- For each $s^* \in (0, \infty)$ there exists $\rho^* : (-1, 1) \to \mathbb{R}$ (a re-parameterization) which is continuous, injective, and

$$\rho^*(0) = s^*, \ \ t \mapsto (\Lambda(\rho^*(t)), \kappa(\rho^*(t))), \ \ t \in (-1, 1), \text{ is analytic.}$$

  This does not imply that $\mathfrak{R}$ is locally a smooth curve. (The map $\sigma : (-1, 1) \to \mathbb{R}^2$ given by $\sigma(t) = (t^2, t^3)$ is real-analytic and its image is a curve with a cusp.) Nor does it preclude the possibility of secondary bifurcation points on $\mathfrak{R}$. In particular, since $(\Lambda, \kappa) : [0, \infty) \to \mathbb{R} \times X$ is not required to be globally injective; self-intersection of $\mathfrak{R}$ (as in a figure eight) is not ruled out.
- Secondary bifurcation points on the bifurcating branch, if any, are isolated.

See Theorem 9.1.1 for a complete statement and §9.3 for an application to the elastic-rod problem. This result about real-analytic global bifurcation from a simple eigenvalue is a sharpened version of a theorem due to Dancer. His general results [24, 26] deal with bifurcation from eigenvalues of higher multiplicity and give the path-connectedness of solutions sets that are not essentially one-dimensional. Since his hypotheses are less restrictive, his conclusions are necessarily somewhat less precise. The topological theory of global bifurcation without analyticity assumptions was developed slightly earlier, first for nonlinear Sturm-Liouville problems (such as (1.3)) by Crandall & Rabinowitz [21], then for partial differential equations by Rabinowitz [50], and for problems with a positivity structure by Dancer [25] and Turner [64]. Their basic tool was an infinite-dimensional topological degree function and the outcome was the existence of a global connected (but not always path-connected) set of solutions. Although it is sometimes possible to argue from the implicit function theorem that the connected set given by topological methods is a smooth curve in $\mathbb{R} \times X$, this approach fails if there is a secondary bifurcation point on the bifurcating branch.

What is important is that in the analytic case a one-dimensional branch can be followed unambiguously through a secondary bifurcation point. In fact a one-dimensional branch is uniquely determined globally by its behaviour in an open set and can be parameterized globally, even when it intersects manifolds of solutions of different dimensions (see §7.5).

1.4 LAYOUT

We begin in Chapter 2 with a review, without proofs, of the linear functional analysis needed for nonlinear theory. Chapter 3 introduces the main results from nonlinear analysis, including the inverse and implicit function theorems for functions of limited differentiability in Banach spaces. Chapter 4 covers similar ground for analytic operators and operator equations in Banach spaces. In Chapters 5, 6 and 7 we consider finite-dimensional analyticity with particular regard to analyticity over the field $\mathbb{R}$. We prove the classical theorems of Weierstrass on the reduction of an analytic equation to a canonical form which involves a polynomial equation for one variable in which the coefficients are analytic functions of the other variables.

Chapter 8 deals with the finite-dimensional reduction of infinite dimensional problems. When the infinite-dimensional problem involves analytic operators, so does the finite-dimensional reduction and the mapping from solutions of the latter to solutions of the former is also analytic. This chapter is the link between the theory of finite-dimensional analytic varieties and infinite-dimensional problems in Banach spaces. Chapter 9 considers what conclusions can then be drawn about global one-dimensional branches of solutions of real-analytic operator equations. This concludes the abstract theory.

Chapter 10 illustrates our discussion of global real-analytic bifurcation theory with a substantial example from mathematical hydrodynamics: the existence question for steady two-dimensional periodic waves on an infinitely deep ocean. There is only one real parameter λ in the problem, the square of the Froude number which represents the speed of the wave.

In his 1847 paper [56] Stokes discussed nonlinear waves with small amplitudes using power series. At the time the proof of convergence was very difficult and only in the 1920s did Nekrasov [47] and Levi-Civita [42], independently, settle the question. Nowadays the existence of small-amplitude water waves can be recognised as nothing more complicated than bifurcation from a simple eigenvalue.

In an 1880 note, Stokes [57] conjectured the existence of a large amplitude periodic wave with a stagnation point and a corner containing an angle of 120° at its highest point. He further speculated that this wave of extreme form marks the limit of steady periodic waves in terms of amplitude (*the Stokes wave of greatest height*).

In Chapter 10 we show how real-analytic global bifurcation theory can account for the existence of waves of all amplitudes from zero up to that of Stokes' highest wave. See [60] for an account of topological methods applied to the same problem; the conclusions there are, in general, weaker.

Almost all the material here is to be found in the literature. The novelty is in the selection and organization of the material with bifurcation theory in mind. Each chapter ends with notes on sources.

PART 1
Linear and Nonlinear Functional Analysis

Chapter Two

Linear Functional Analysis

In this chapter we introduce notation and record, without proof, the main results from linear functional analysis used in the sequel.

2.1 PRELIMINARIES AND NOTATION

In what follows

$\mathbb{R}$ denotes the field of real numbers,

$\mathbb{C}$ denotes the field of complex numbers,

$\mathbb{N}$ denotes the natural numbers, not including 0; $\mathbb{N}_0 = \mathbb{N} \cup \{0\}$,

$\mathbb{Z}$ denotes the integers,

$\mathbb{F}$ denotes $\mathbb{R}$ or $\mathbb{C}$ in statements which are true for both,

$\mathfrak{Re}\, z, \mathfrak{Im}\, z, \bar{z}$ and $|z|$ denote the real part, the imaginary part, the complex conjugate and the modulus, respectively, of $z \in \mathbb{C}$,

S_k denotes the symmetric group of permutations of $\{1, \cdots, k\}$.

We will assume familiarity with concepts such as closedness, completeness, compactness and connectedness in metric spaces, and with linearity, linear independence and dimension in vector spaces over $\mathbb{R}$ and $\mathbb{C}$. A maximal connected subset of a metric space M is called a component of M.

A subset C of a linear space X over $\mathbb{F}$ is said to be convex if for all $x_1, x_2 \in C$ and every $t \in [0, 1]$, $(1 - t)x_1 + tx_2 \in C$. In other words C contains every line-segment joining any two points of C.

A norm on a linear space X over $\mathbb{F}$ is an $\mathbb{R}$-valued function $\| \cdot \|$ such that

$$\|x\| \geq 0 \text{ for all } x \in X \text{ with equality if and only if } x = 0,$$
$$\|\alpha x\| = |\alpha| \, \|x\| \text{ for all } x \in X \text{ and } \alpha \in \mathbb{F},$$
$$\|x + y\| \leq \|x\| + \|y\| \text{ for all } x, \, y \in X.$$

Various symbols, for example $\| \cdot \|$, $\| \cdot \|_X$ or $\|| \cdot \||$, could be used to denote norms on a space X, but when the intention is clear from the context, we use the generic symbol $\| \cdot \|$ in all cases.

Two norms $\|| \cdot \||$ and $\| \cdot \|$ on the same linear space X are said to be equivalent if there exist positive constants k, K, such that

$$k\|x\| \leq \||x\|| \leq K\|x\| \text{ for all } x \in X.$$

If $\| \cdot \|$ is a norm on a linear space X,

$$\rho(x,y) = \|x - y\|, \quad x, y \in X,$$

defines a metric ρ on X. If the metric space (X, ρ) is complete, $(X, \| \cdot \|)$ is called a Banach space. (A Banach space is said to be real or complex if the field $\mathbb{F}$ is $\mathbb{R}$ or $\mathbb{C}$, but when the field does not matter it will not be mentioned.)

If a sequence $\{x_n\}$ in a Banach space X has $\sum_{n \in \mathbb{N}} \|x_n\| < \infty$ then the sequence $\left\{ \sum_{k=1}^{n} x_k \right\}$ of partial sums is Cauchy, and hence convergent, in X. We then say that the series $\left\{ \sum_{k=1}^{\infty} x_k \right\}$ is summable in norm, which implies that it is convergent to the same sum irrespective of rearrangement of its terms.

LEMMA 2.1.1 *Every finite-dimensional linear space with any norm is a Banach space. Any two norms on a finite-dimensional linear space are equivalent.*

EXAMPLES 2.1.2 (a) The finite-dimensional space $\mathbb{F}^N$, $N \in \mathbb{N}$, is a Banach space over $\mathbb{F}$ with norm $| \cdot |$ where

$$|(z_1, z_2, \cdots, z_N)|^2 = \sum_{k=1}^{N} |z_k|^2, \quad (z_1, \cdots, z_N) \in \mathbb{F}^N.$$

(b) The infinite-dimensional linear space $C^n([0,1], \mathbb{F}^N)$ of continuous functions $u = (u_1, \cdots, u_N) : [0,1] \to \mathbb{F}^N$, the nth derivatives of which exist on $(0,1)$ and have continuous extensions to $[0,1]$, with the norm

$$\|u\| = \sup \{|u(x)| + |u^{(n)}(x)| : x \in (0,1)\},$$

where

$$u^{(n)} = \Big(\frac{d^n u_1}{dx^n}, \cdots, \frac{d^n u_N}{dx^n}\Big), \quad u = (u_1, \cdots, u_N),$$

is a Banach space. By convention, $C([0,1], \mathbb{F}^N) = C^0([0,1], \mathbb{F}^N)$. □

Other Banach spaces, in particular Sobolev spaces of functions, are introduced in [10] and [32]. The following result is due to F. Riesz.

LEMMA 2.1.3 *A Banach space X is finite-dimensional if and only if the closed unit ball $\{x \in X : \|x\| \le 1\}$ is compact.*

Both the closed unit ball and the open unit ball $\{x \in X : \|x\| < 1\}$ are convex. The following is a corollary of the open mapping theorem (2.4.2 below).

LEMMA 2.1.4 *Suppose that $\| \cdot \|$ and $||| \cdot |||$ are two Banach-space norms on the linear space X and that $|||x||| \le K\|x\|$ for all $x \in X$. Then $||| \cdot |||$ and $\| \cdot \|$ are equivalent norms on X.*

An inner product $\langle \cdot, \cdot \rangle$ on a linear space X over $\mathbb{F}$ is an $\mathbb{F}$-valued function on $X \times X$ such that

$$y \mapsto \langle x, y \rangle \text{ is } \mathbb{F}\text{-linear for each (fixed) } x \in X,$$

$$\langle x, y \rangle = \overline{\langle y, x \rangle} \text{ for all } (x, y) \in X \times X,$$

$$\langle x, x \rangle \geq 0 \text{ for all } x \in X \text{ with equality if and only if } x = 0.$$

An inner product $\langle \cdot, \cdot \rangle$ is linear with respect to each of the variables separately if $\mathbb{F} = \mathbb{R}$, but only with respect to the second variable if $\mathbb{F} = \mathbb{C}$. If $\langle \cdot, \cdot \rangle$ is an inner product on X then

$$\|x\|^2 = \langle x, x \rangle, \quad x \in X,$$

defines a norm on X and the Cauchy-Schwarz inequality,

$$|\langle x, y \rangle| \leq \|x\| \|y\| \text{ for all } x, y \in X,$$

holds. A complete inner-product space is called a Hilbert space. It is clear that the Banach space $\mathbb{F}^N$ defined above is a Hilbert space over $\mathbb{F}$ because

$$\langle (z_1, \cdots, z_N), (\hat{z}_1, \cdots, \hat{z}_N) \rangle = \sum_{k=1}^{N} \overline{z_k} \, \hat{z}_k$$

is an inner-product on $\mathbb{F}^N$ over $\mathbb{F}$.

2.2 SUBSPACES

Suppose that X is a Banach space with norm $\| \cdot \|$ and that Y is a linear subspace of X. Then $(Y, \| \cdot \|)$ is a Banach space if and only if Y is closed in X. If X_1, X_2 are closed linear subspaces of a Banach space X with the property that $X_1 \cap X_2 = \{0\}$ and, for every $x \in X$, there exist $x_1 \in X_1$ and $x_2 \in X_2$ such that $x = x_1 + x_2$, we say that X is the topological direct sum of X_1 and X_2 and write $X = X_1 \oplus X_2$. (This notation always has the implication that X_1 and X_2 are Banach spaces with the norm inherited from X.) The space X_2 is called the topological complement of X_1 in X, and *vice versa*.

LEMMA 2.2.1 *If* $X = X_1 \oplus X_2$, *then every* $x \in X$ *can be written in a unique way as* $x = x_1 + x_2$, $x_i \in X_i$, $i = 1, 2$.

Even when X_1 is closed it is not always true that $X = X_1 \oplus X_2$ for some closed linear subspace X_2; X_1 may not have a topological complement. However when X_1 is finite-dimensional we have the following.

LEMMA 2.2.2 *Suppose that* X *is a Banach space and* X_1 *is a finite-dimensional subspace of* X. *Then* X_1 *is closed and there exists a closed subspace* X_2 *of* X *such that* $X = X_1 \oplus X_2$. *In this case* $\dim X_1$ *is called the codimension of* X_2.

2.3 DUAL SPACES

Suppose that X is a Banach space over $\mathbb{F}$. A linear functional f on X is a linear mapping from X into $\mathbb{F}$. A linear functional f is said to be bounded if there exists K such that

$$|f(x)| \leq K\|x\| \text{ for all } x \in X.$$

The set of all bounded linear functionals on X is a linear space over $\mathbb{F}$ which is usually denoted by X^*. It becomes a Banach space when endowed with the norm defined by

$$\|f\| = \inf\{K \geq 0 : |f(x)| \leq K\|x\| \text{ for all } x \in X\}.$$

The Banach space X^* with this norm is called the dual space of X. An important corollary of the Hahn-Banach theorem is the following.

LEMMA 2.3.1 *Let X be a Banach space.*
(a) For each $x \in X$ there exists $f_x \in X^*$ such that $f_x(x) = \|x\|$ and $\|f_x\| = 1$.
(b) Suppose that E is a closed linear subspace of X of codimension 1. Then there exists $f \in X^*$ such that $f(x) = 0$ if and only if $x \in E$. In other words, if $x_0 \neq 0$ and $X = \text{span}\{x_0\} \oplus E$ there exists $f \in X^*$ such that $f(x_0) = 1$ and $\ker f = E$.

Note that in *(a)* f_x need not be unique. For example, let X denote $\mathbb{R}^2$ with the norm $\|(x_1, x_2)\| = \max\{|x_1|, |x_2|\}$. For $t \in [0, 1]$ let $f_{(t)}(x_1, x_2) = tx_1 + (1 - t)x_2$. Then

$$|f_{(t)}(x_1, x_2)| = |tx_1 + (1 - t)x_2| \leq t|x_1| + (1 - t)|x_2| \leq \|(x_1, x_2)\|,$$

whence $\|f_{(t)}\| \leq 1$. Since $\|(1, 1)\| = 1 = f_{(t)}(1, 1)$ for all $t \in [0, 1]$, f_x in Lemma 2.3.1 is not unique when $x = (1, 1) \in X$.

EXAMPLES 2.3.2 (a) Let $(X, \langle \cdot, \rangle)$ be a Hilbert space. For any $x \in X$ with $x \neq 0$, the functional defined by

$$f_x(y) = \frac{\langle x, y \rangle}{\|x\|}$$

satisfies the requirements of Lemma 2.3.1 (a).

(b) Let X denote the Banach space $C([0, 1], \mathbb{F}^N)$. For $u \in X \setminus \{0\}$, choose $t_u \in [0, 1]$ and $\theta_j \in [-\pi, \pi]$ such that

$$\|u\|^2 = \sum_{j=1}^{N} u_j^2(t_u)e^{i\theta_j}.$$

Now let

$$f_u(v) = \|u\|^{-1} \sum_{j=1}^{N} u_j(t_u)v_j(t_u)e^{i\theta_j} \text{ for } v \in X.$$

Then $f_u \in X^*$, $\|f_u\| = 1$ and $f_u(u) = \|u\|$, as in Lemma 2.3.1 (a). $\qquad\square$

It is clear that if X is a Hilbert space then, for any $z \in X$, an element of $f \in X^*$ can be defined by $f(x) = \langle z, x \rangle$ for all $x \in X$. That the converse is also true is the content of the next theorem.

THEOREM 2.3.3   (Riesz Representation Theorem) *Let X be a Hilbert space and $f \in X^*$. Then there exists a unique $x_f \in X$ such that*

$$f(x) = \langle x_f, x \rangle \text{ and } \|x\| = \|x_f\| \text{ for all } x \in X.$$

2.4 LINEAR OPERATORS

Before discussing the boundedness of linear operators between Banach spaces we pause to note that the familiar notion of linearity is intimately connected with the field $\mathbb{F}$. For example, let $\mathbb{C}_r$ and $\mathbb{C}_c$ denote the complex numbers regarded as a linear space over $\mathbb{R}$, and over $\mathbb{C}$, respectively. (The spaces are the same, only the fields are different.) Now consider the operator L on $\mathbb{C}$ defined by $Lz = \bar{z}$, the operation of complex conjugation. This is a linear operator on $\mathbb{C}_r$, but *not* on $\mathbb{C}_c$, an observation which will be important when we consider the Fréchet differentiability and analyticity of nonlinear operators.

Suppose that X, Y are Banach spaces over the same field $\mathbb{F}$. A linear function $A : X \to Y$ is said to be a bounded linear operator, written $A \in \mathcal{L}(X, Y)$, if there exists $K \geq 0$ such that

$$\|Ax\| \leq K\|x\| \text{ for all } x \in X.$$

An important element of $\mathcal{L}(X, X)$ for any Banach space X is the identity operator I, defined by $Ix = x$ for all $x \in X$. An obvious element of $\mathcal{L}(X, Y)$ is the zero operator, defined by $0x = 0 \in Y$ for all $x \in X$.

It is easily seen that a linear mapping $A : X \to Y$ is continuous if and only if it is bounded. It is also clear that when X, Y are Banach spaces over $\mathbb{F}$, $\mathcal{L}(X, Y)$ is a linear space over $\mathbb{F}$ with the natural definitions of addition and multiplication by scalars. Even more is true.

LEMMA 2.4.1   *If X, Y are Banach spaces over $\mathbb{F}$, then $\mathcal{L}(X, Y)$ is a Banach space when endowed with the norm*

$$\|B\| = \inf\{K \geq 0 : \|Bx\| \leq K\|x\| \text{ for all } x \in X\}.$$

In particular, $\mathcal{L}(X, \mathbb{F}) = X^*$. For $A \in \mathcal{L}(X, Y)$ let

$$\ker(A) = \{x \in X : Ax = 0 \in Y\},$$
$$\text{range}(A) = \{y \in Y : y = Ax \text{ for some } x \in X\}.$$

Both $\ker(A) \subset X$ and range $(A) \subset Y$ are linear spaces and $\ker(A)$ is closed because A is continuous. However range (A) need not be closed and for nonlinear analysis it will be important to have hypotheses on A sufficient to guarantee that it is, see §2.7. An operator $A \in \mathcal{L}(X, Y)$ is injective if $\ker(A) = \{0\}$, surjective if range $(A) = Y$ and bijective if it is both. An operator $B \in \mathcal{L}(X, Y)$ is said to be a homeomorphism if it is a bijection and $B^{-1} \in \mathcal{L}(Y, X)$.

THEOREM 2.4.2 **(Open Mapping Theorem)** *Suppose X and Y are Banach spaces and that $B \in \mathcal{L}(X, Y)$ is surjective. Then $B(U)$ is open in Y for all open sets $U \subset X$.*

We have already seen an important corollary of the open mapping theorem, Lemma 2.1.4. Here is another.

COROLLARY 2.4.3 **(Corollary of Open Mapping Theorem)** *If X and Y are Banach spaces and $B \in \mathcal{L}(X, Y)$ is a bijection, then B is a homeomorphism.*

This result will be important when we have to check the hypotheses of the implicit or inverse function theorems which require a certain operator to be a homeomorphism between Banach spaces. It shows that being bijective and bounded is enough. For $A \in \mathcal{L}(X, X)$ let $A^k = \underbrace{A \circ A \circ \cdots \circ A}_{k \text{ times}}$.

2.5 NEUMANN SERIES

Let $A \in \mathcal{L}(X, X)$ be such that $\|A\| < 1$. Then the sequence of partial sums $\sum_{k=0}^{N} A^k$ is a Cauchy sequence which converges in the Banach space $\mathcal{L}(X, X)$ and the limit is denoted by $\sum_{k=0}^{\infty} A^k$. It is easily seen that

$$(I - A)\left(\sum_{k=0}^{\infty} A^k\right) = I = \left(\sum_{k=0}^{\infty} A^k\right)(I - A), \quad (I - A)^{-1} = \sum_{k=0}^{\infty} A^k, \quad (2.1)$$

and $I - A$ is a homeomorphism. The series on the right is called the Neumann series of A. Note that because

$$\left\|\sum_{k=0}^{\infty} A^k\right\| \leq \sum_{k=0}^{\infty} \|A\|^k < \infty,$$

the series in (2.1) is summable in norm.

Now suppose that $T \in \mathcal{L}(X, Y)$ is a homeomorphism. If $S \in \mathcal{L}(X, Y)$ is such that $\|T^{-1}(S - T)\| < 1$, it follows that $I + T^{-1}(S - T) : X \to X$ has a bounded inverse given by a Neumann series. Since $S = T(I + T^{-1}(S - T))$, the existence of $S^{-1} \in \mathcal{L}(Y, X)$, and hence the following result, is now immediate.

LEMMA 2.5.1 *If $T \in \mathcal{L}(X, Y)$ is a homeomorphism and $S \in \mathcal{L}(X, Y)$ has $\|S - T\| < \|T^{-1}\|^{-1}$, then S is a homeomorphism. In fact the homeomorphisms form an open set in $\mathcal{L}(X, Y)$ on which the mapping $S \mapsto S^{-1}$ is continuous from $\mathcal{L}(X, Y)$ to $\mathcal{L}(Y, X)$.*

If $A \in \mathcal{L}(X, Y)$ then $A^* \in \mathcal{L}(Y^*, X^*)$ is the linear operator, called the conjugate of A, defined by

$$(A^* f)(x) = f(A(x)) \text{ for all } x \in X \text{ and } f \in Y^*.$$

2.6 PROJECTIONS AND SUBSPACES

Suppose that X is a Banach space and that $P \in \mathcal{L}(X, X)$ has the property that

$$P(Px) = Px \text{ for all } x \in X.$$

Then P is said to be a projection or a projection operator. (In particular, projection operators are bounded.) Note that P is a projection if and only if $I - P$ is a projection. Clearly the zero operator and the identity on X are projections. The connection between projection operators and topological direct sums is summarized in the following two propositions.

PROPOSITION 2.6.1 *Suppose that X is a Banach space and that $P : X \to X$ is a projection. Then*

$$\ker(P) \text{ and } \operatorname{range}(P) \text{ are closed subspaces of } X;$$
$$\operatorname{range}(P) = \ker(I - P) \text{ and } \ker(I - P) = \operatorname{range}(P);$$
$$\ker(P) \oplus \operatorname{range}(P) = X.$$

PROPOSITION 2.6.2 *Suppose that X is a Banach space with $X = X_1 \oplus X_2$. By Lemma 2.2.1, $x \in X$ can be written in a unique way as $x = x_1 + x_2$ where $x_1 \in X_1$ and $x_2 \in X_2$. Let $Px = x_1$, so that $(I - P)x = x_2$. Then both P and $I - P$ are projections on X. P is called the projection onto X_1 parallel to X_2.*

That P and $I - P$ defined in the second proposition are linear follows easily from the definition of a topological direct sum. What is not so obvious is that P, so defined, is a bounded operator. This is yet another corollary of the open mapping theorem 2.4.2. Boundedness of projections is essential for our purposes. Note that a closed subspace X_1 of X alone does not define a projection and there is no such notion as "the projection onto X_1". Indeed, if X_1 has no topological complement, then there does not exist a projection from X onto X_1. However, because of Lemma 2.2.2 we have the following.

LEMMA 2.6.3 *If X_1 is a finite-dimensional subspace of a Banach space X, then there exists a projection P on X with $X_1 = \operatorname{range}(P)$.*

DEFINITION 2.6.4 *Suppose $X_1, \cdots, X_n$ are Banach spaces over a field $\mathbb{F}$. Then the product space of n-tuples $(x_1, \cdots, x_n)$, $x_i \in X_i$, is a Banach space denoted by $X_1 \times \cdots \times X_n$. Many equivalent norms may be defined on $X_1 \times \cdots \times X_n$, but we shall assume that*

$$\|(x_1, \cdots, x_n)\| = \sqrt{\sum_{k=1}^{n} \|x_k\|^2}.$$

When each of the spaces X_k is a Hilbert space, the product space is also a Hilbert space with inner product

$$\langle (x_1, \cdots, x_n), (z_1, \cdots z_n) \rangle = \sum_{k=1}^{n} \langle x_k, z_k \rangle,$$

for $(x_1, \cdots, x_n)$, $(z_1, \cdots, z_n) \in X_1 \times \cdots \times X_n$.

From time to time it will be convenient to identify the space X_k with

$$\underbrace{\{0\} \times \cdots \times \{0\}}_{k-1 \text{ terms}} \times X_k \times \underbrace{\{0\} \times \cdots \times \{0\}}_{n-k \text{ terms}} \subset X_1 \times \cdots \times X_n,$$

and hence with a closed subspace of the product space. There is an obvious projection of the product space onto the space identified with X_k, parallel to the product of the other spaces.

2.7 COMPACT AND FREDHOLM OPERATORS

Suppose that (M, ρ) is a compact metric space. Let $C(M, \mathbb{F})$ denote the linear space of continuous functions $u : M \to \mathbb{F}$ with the norm

$$\|u\| = \max\{|u(x)| : x \in M\}.$$

Then $C(M, \mathbb{F})$ is a Banach space. The following characterizes its compact sets.

THEOREM 2.7.1   (Ascoli-Arzelà Theorem) *Suppose that M is a compact metric space. A set $B \subset C(M, \mathbb{F})$ has compact closure if and only if (i) there exists k such that $\|u\| \le k$ for all $u \in B$, and (ii) given $\epsilon > 0$ there exists $\delta > 0$ (independent of u) such that for all $u \in B$ $|u(x) - u(y)| < \epsilon$ if $\rho(x, y) < \delta$.*

Let X and Y be Banach spaces. An operator $K \in \mathcal{L}(X, Y)$ is said to be compact if every bounded sequence $\{x_n\} \subset X$ has a subsequence $\{x_{n_k}\}$ for which $\{K(x_{n_k})\}$ converges in Y. In finite dimensions, all linear operators are bounded, and therefore compact, by Lemma 2.1.3. As a consequence, if Y is finite dimensional and X is a Banach space, then any linear operator $L : Y \to X$ is compact and any $A \in \mathcal{L}(X, Y)$ is compact. Note that if W, X, Y and Z are Banach spaces and $K \in \mathcal{L}(X, Y)$ is compact, $B \in \mathcal{L}(Z, X)$ and $C \in \mathcal{L}(Y, W)$, then $C \circ K \circ B \in \mathcal{L}(Z, W)$ is compact.

Theorem 2.7.1 leads to an important class of compact operators.

DEFINITION 2.7.2   *Suppose that X and Y are linear spaces over $\mathbb{F}$ and $X \subset Y$. Now suppose that $(X, |\cdot|)$ and $(Y, \|\cdot\|)$ are Banach spaces for which the mapping $\iota : X \to Y$, given by $\iota(x) = x \in Y$ for all $x \in X$, is bounded. We say that the embedding of X in Y is continuous and ι is called the embedding operator. If ι is compact, we say that X is compactly embedded in Y.*

EXAMPLE 2.7.3   The prototypical example of compact embeddings is the following. Let $X = C^1([0,1], \mathbb{F})$, $Y = C([0,1], \mathbb{F})$, defined in Example 2.1.2 (b), and let $Kf = f \in Y$ for $f \in X$. Then K is compact, by the Ascoli-Arzelà theorem and the mean-value theorem for functions of one variable. In this example K coincides with the embedding operator ι from $X \subset Y$ into Y. $\square$

Another source of compact operators is the following.

LEMMA 2.7.4 *When X and Y are Banach spaces the compact operators form a closed linear subspace of $\mathcal{L}(X,Y)$. In particular, if there exists a sequence $\{A_n\} \subset \mathcal{L}(X,Y)$ such that $\|A_n - A\| \to 0$ as $n \to \infty$ and A_n has finite-dimensional range for each n, then A is a compact operator.*

An operator $A \in \mathcal{L}(X,Y)$ is a Fredholm operator of index p if

$\ker(A)$ has finite dimension n;

range (A) is closed and has finite codimension r;

$p = n - r$.

Clearly any homeomorphism from X to Y is a Fredholm operator of index zero. The following result for compact operators in Banach spaces contains the dimension theorem in linear algebra as a special case and explains the central rôle of compact operators in functional analysis.

THEOREM 2.7.5   (Fredholm Alternative) *Let $K \in \mathcal{L}(X,X)$ be compact. Then $I - K \in \mathcal{L}(X,X)$ and $I - K^* \in \mathcal{L}(X^*,X^*)$ are Fredholm operators of index zero. Moreover*

$$\dim \ker(I - K) = \dim \ker(I - K^*)$$
$$= \operatorname{codim\,range} (I - K) = \operatorname{codim\,range} (I - K^*).$$

The following criterion ensures that an operator $B \in \mathcal{L}(X,Y)$, for different spaces X, Y, is Fredholm in an infinite dimensional setting.

THEOREM 2.7.6 *Suppose X and Y are Banach spaces, $K \in \mathcal{L}(X,Y)$ is compact and $T \in \mathcal{L}(X,Y)$ is a homeomorphism. Then $B = T + K$ is Fredholm with index zero.*

Proof. It suffices to notice that $B = T + K = T(I + T^{-1}K)$ and, since T, T^{-1} are bounded, $B \in \mathcal{L}(X,Y)$ is Fredholm with index zero if and only if $I + T^{-1}K \in \mathcal{L}(X,X)$ is Fredholm with index zero. However $T^{-1}K \in \mathcal{L}(X,X)$ is compact, and the result follows from Theorem 2.7.5. $\qquad\qquad\qquad\square$

PROPOSITION 2.7.7 *[67] Let X, Y be Banach spaces. Then the set of Fredholm operators is an open set in $\mathcal{L}(X,Y)$ and the Fredholm index of operators is constant on the components of this set.*

DEFINITION 2.7.8 *Let $\iota \in \mathcal{L}(X,Y)$ denote the continuous embedding of a Banach space X in a Banach space Y. Suppose that $\lambda_0 \in \mathbb{F}$ and $A \in \mathcal{L}(X,Y)$ are such that $\lambda_0\iota - A$ is Fredholm of index zero, that $\ker(\lambda_0\iota - A)$ is one-dimensional over $\mathbb{F}$, and that range $(\lambda_0\iota - A) \cap \iota\big(\ker(\lambda_0\iota - A)\big) = \{0\}$. Then we say that λ_0 is a simple eigenvalue of A. (When $X = Y$ is finite-dimensional, this is equivalent to saying that λ_0 is an eigenvalue of A of algebraic multiplicity 1.) An element $\xi_0 \in X \setminus \{0\}$ with $A\xi_0 = \lambda_0\iota\,\xi_0$ is called an eigenvector of A corresponding to the eigenvalue λ_0.*

LEMMA 2.7.9 *Suppose that λ_0 is a simple eigenvalue of A with eigenvector ξ_0. Let*

$$Y_1 = \text{range} \, (\lambda_0 \iota - A) \text{ and } X_1 = \iota^{-1}(Y_1)\big(= X \cap Y_1\big).$$

Then $X = X_1 \oplus \text{span} \, \{\xi_0\}$ and $\lambda_0 \iota - A$ is a homeomorphism from X_1 to Y_1 with the norms inherited from X and Y.

Proof. For any $y \in Y$, $y = \alpha \iota \xi_0 + y_1$, $y_1 \in Y_1$, $\alpha \in \mathbb{F}$, since Y_1 has codimension 1 and $\iota \xi_0 \notin Y_1$. In particular, for $x \in X$, $\iota x = \alpha \iota \xi_0 + y_1$ and $x - \alpha \xi_0 \in \iota^{-1}(Y_1) = X_1$. Hence $x = x_1 + \alpha \iota \xi_0$ for some $x_1 \in X_1$ and $\alpha \in \mathbb{F}$.

Next, note that $X_1 = \iota^{-1}(Y_1)$ is closed in X since Y_1 is closed in Y and $\iota : X \to Y$ is continuous. Since $X_1 \cap \text{span} \, \{\xi_0\} = \iota^{-1}\big(Y_1 \cap \text{span} \, \{\iota \xi_0\}\big) = \{0\}$, it follows that $X = X_1 \oplus \text{span} \, \{\xi_0\}$. So $\lambda_0 \iota - A \in \mathcal{L}(X_1, Y_1)$ is a bijection, and hence a homeomorphism.

This completes the proof. □

2.8 NOTES ON SOURCES

This is standard material, to be found, for example, in the books of Brezis [10], Friedman [32], Kreyszig [41], Rudin [51] or Taylor [58]. The theory of Fredholm operators and their indices is covered in the books by Kato [35] and Wloka [67].

Chapter Three

Calculus in Banach Spaces

We turn to our main topic, nonlinear operators between Banach spaces.

"Big O" and "little o" notation. Suppose that f and g are functions defined from a neighbourhood of a in a Banach space X to a Banach space Y. We write

$$f(x) = g(x) + o(x - a) \text{ as } x \to a \text{ if } \lim_{x \to a} \frac{\|f(x) - g(x)\|}{\|x - a\|} = 0,$$

and

$$f(x) = g(x) + O(x - a) \text{ as } x \to a \text{ if } \limsup_{x \to a} \frac{\|f(x) - g(x)\|}{\|x - a\|} < \infty.$$

3.1 FRÉCHET DIFFERENTIATION

Suppose that X and Y are Banach spaces, $U \subset X$ is open, and $F : U \to Y$.

DEFINITION 3.1.1 *The function F is Fréchet differentiable at $x_0 \in U$ if there exists $A \in \mathcal{L}(X, Y)$ such that*

$$\lim_{0 < \|h\| \to 0} \frac{\|F(x_0 + h) - F(x_0) - Ah\|}{\|h\|} = 0.$$

If such an operator A exists, it is unique and is called the Fréchet derivative of F at x_0. We write $A = dF[x_0]$. The evaluation $dF[x_0]x \in Y$, for any $x \in X$, is called the directional derivative of F at x_0 in the direction x. It is said that F is Fréchet differentiable on U if it is Fréchet differentiable at every point of U, in which case $x \mapsto dF[x]$ is a function from U to $\mathcal{L}(X, Y)$ which we denote by dF.

An equivalent way of saying that $A \in \mathcal{L}(X, Y)$ is the derivative of F at x_0 is to write

$$F(x_0 + h) - F(x_0) - Ah = o(\|h\|) \text{ as } h \to 0.$$

Obviously F is continuous at x_0 if it is Fréchet differentiable at x_0.

REMARK 3.1.2 A Fréchet derivative belongs to neither X nor Y, but rather is a bounded linear operator from X to Y. (To say that $\cos x_0$ is the derivative at x_0 of the function $f : \mathbb{R} \to \mathbb{R}$ given by $f(x) = \sin x$ means only that $df[x_0]x = x \cos x_0$ for all $x \in \mathbb{R}$.) $\qquad\square$

Two operations occur repeatedly in the manipulation of derivatives: addition and composition.

LEMMA 3.1.3 **(Addition)** *Suppose that* $F, G : U \to Y$. *If* $dF[x_0]$ *and* $dG[x_0]$ *both exist, then* $d(F + G)[x_0]$ *exists and*

$$d(F + G)[x_0] = dF[x_0] + dG[x_0] \in \mathcal{L}(X, Y).$$

(Chain Rule) *Let* X, Y *and* Z *be Banach spaces,* U, V *open sets and suppose that* $F : U(\subset X) \to Y$ *and* $G : V(\subset Y) \to Z$ *are such that* $G \circ F : U \to Z$ *is defined,* $dF[x_0]$ *exists and* $dG[F(x_0)]$ *exists. Then* $d(G \circ F)[x_0]$ *exists and*

$$d(G \circ F)[x_0] = dG[F(x_0)] \circ dF[x_0] \in \mathcal{L}(X, Z).$$

Proof. The proof of the first part is elementary and we leave it as an exercise.
Suppose $h \neq 0$. Then for $x_0 \in U$ and $h \in X$ with $\|h\|$ sufficiently small,

$$
\begin{aligned}
G(F(x_0 + h)) &- G(F(x_0)) \\
&= G\big(F(x_0) + (F(x_0 + h) - F(x_0))\big) - G(F(x_0)) \\
&= dG[F(x_0)](F(x_0 + h) - F(x_0)) + o(\|F(x_0 + h) - F(x_0)\|)
\end{aligned}
$$

as $\|h\| \to 0$, since F is continuous at x_0. Now

$$F(x_0 + h) - F(x_0) = dF[x_0]h + o(\|h\|) \text{ as } \|h\| \to 0,$$

and so

$$
\begin{aligned}
G(F(x_0 + h)) &- G(F(x_0)) \\
&= dG[F(x_0)](F(x_0 + h) - F(x_0)) + o(\|h\|)) \\
&= dG[F(x_0)]dF[x_0](h) + o(\|h\|) \text{ as } \|h\| \to 0,
\end{aligned}
$$

since $dG[F(x_0)]$ is linear and $F(x_0 + h) - F(x_0) - dF[x_0]h = o(\|h\|)$ as $h \to 0$. Therefore

$$\frac{\|G(F(x_0 + h)) - G(F(x_0)) - dG[F(x_0)]dF[x_0](h)\|}{\|h\|}$$

$$= \frac{o(\|h\|)}{\|h\|} \to 0 \text{ as } \|h\| \to 0.$$

This proves the result. $\qquad\qquad\qquad\qquad\qquad\qquad\qquad\qquad\qquad\qquad\square$

The following corollary of the chain rule is a Banach-space substitute for the mean-value theorem from the theory of functions of one variable.

LEMMA 3.1.4 *Let* X *and* Y *be Banach spaces,* $U \subset X$ *a convex open set, and let* $F : U \to Y$ *be Fréchet differentiable at each point of* U *with*

$$\sup\{\|dF[x]\| : x \in U\} = m < \infty.$$

Then

$$\|F(x_1) - F(x_2)\| \leq m\|x_1 - x_2\|, \quad x_1, x_2 \in U.$$

Proof. Choose and fix x_1, $x_2 \in U$ and let $x(t) = (t-1)x_2 + (2-t)x_1$, $t \in [1,2]$. Note that $x(t) \in U$ because U is convex. Now use Lemma 2.3.1 (a) to choose $g \in Y^*$ such that

$$g(F(x_2) - F(x_1)) = \|F(x_2) - F(x_1)\| \text{ with } \|g\| = 1$$

and define $u : [1,2] \to \mathbb{R}$ by

$$u(t) = \Re\, g\big(F(x(t))\big).$$

Then u is continuous on $[1,2]$ and, by the chain rule, differentiable on $(1,2)$ with

$$u'(t) = \Re\, g\big(dF[x(t)](x_2 - x_1)\big), \quad t \in (1,2).$$

Since $\|g\| = 1$ it follows that

$$|u'(t)| \leq \|dF[x(t)]\| \, \|x_2 - x_1\| \leq m\|x_2 - x_1\|.$$

Therefore, by the mean-value theorem in one dimension,

$$\|F(x_2) - F(x_1)\| = \Re\, g(F(x(2)) - F(x(1)))$$
$$= |u(2) - u(1)| \leq m\|x_2 - x_1\|.$$

This completes the proof. □

If $\|F(x) - F(y)\| \leq K\|x - y\|$ for all $x, y \in U$ then we say that F is Lipschitz continuous on U with Lipschitz constant K. A mapping with Lipschitz constant $K < 1$ is called a contraction mapping.

DEFINITION 3.1.5 **(Partial Derivatives)** *Suppose that X, Y, and Z are Banach spaces, that $U \subset X \times Y$ is open and $F : U \to Z$. Suppose that $(x_0, y_0) \in U$. Then $U_{x_0} = \{y \in Y : (x_0, y) \in U\}$ is open. If the function $F(\cdot, y_0)$ has a Fréchet derivative at x_0 we denote it by $\partial_x F[(x_0, y_0)] \in \mathcal{L}(X, Z)$ and refer to it as the partial Fréchet derivative of F with respect to x at (x_0, y_0). Similarly $\partial_y F[(x_0, y_0)] : Y \to Z$ will denote the partial Fréchet derivative of F with respect to y.*

EXAMPLE 3.1.6 Let $U = X \times Y = \mathbb{R} \times \mathbb{R}$, let $F(x, y) = 1$ when $xy = 0$ and let $F(x, y) = 0$ otherwise. Clearly $F : \mathbb{R}^2 \to \mathbb{R}$ is not continuous at $(0, 0)$ and is therefore not Fréchet differentiable there. However $\partial_x F[(0, 0)] = \partial_y F[(0, 0)] = 0$. A question naturally arises about the relation between the existence of $dF[(x_0, y_0)]$ and that of $\partial_x F[(x_0, y_0)]$ or $\partial_y F[(x_0, y_0)]$. □

LEMMA 3.1.7 *Suppose that X, Y and Z are Banach spaces, $U \subset X \times Y$ is open.*

(a) If $F : U \to Z$ is such that $dF[(x_0, y_0)]$ exists, then $\partial_x F[(x_0, y_0)]$ and $\partial_y F[(x_0, y_0)]$ exist with, for $(x, y) \in X \times Y$,

$$dF[(x_0, y_0)](x, y) = \partial_x F[(x_0, y_0)]x + \partial_y F[(x_0, y_0)]y.$$

In particular, $\partial_x F[(x_0, y_0)]x \in Z$ is the directional derivative (see Definition 3.1.1) of F at (x_0, y_0) in the direction $(x, 0) \in X \times Y$. If dF is continuous in a neighbourhood of (x_0, y_0), then so are $\partial_x F$ and $\partial_y F$.

(b) Suppose that $\partial_x F[(x_0, y_0)]$ exists and $\partial_y F$, which is defined at every point of a neighbourhood of (x_0, y_0), is continuous at $(x_0, y_0) \in U$. Then $dF[(x_0, y_0)]$ exists.

Proof. (a) Suppose that $dF[(x_0, y_0)]$ exists. Then, by definition,

$$\|F(x_0, y_0 + y) - F(x_0, y_0) - dF[(x_0, y_0)](0, y)\| = o(\|y\|)$$

as $\|y\| \to 0$. Since $\|y\| = \|(0, y))\|$ it follows that

$$\partial_y F[(x_0, y_0)] = dF[(x_0, y_0)](0, \cdot) \in \mathcal{L}(Y, Z).$$

Similarly

$$\partial_x F[(x_0, y_0)] = dF[(x_0, y_0)](\cdot, 0) \in \mathcal{L}(X, Z),$$

the formula in (a) clearly holds, and the continuity of the partial derivatives follows from that of dF.

(b) Note that

$$\|F(x_0 + x, y_0 + y) - F(x_0, y_0) - \partial_x F[(x_0, y_0)]x - \partial_y F[(x_0, y_0)]y\|$$
$$\leq \|F(x_0 + x, y_0 + y) - F(x_0 + x, y_0) - \partial_y F[(x_0, y_0)]y\|$$
$$+ \|F(x_0 + x, y_0) - F(x_0, y_0) - \partial_x F[(x_0, y_0)]x\| = I_1 + I_2, \text{ say.}$$

Now since $\partial_x F[(x_0, y_0)]$ exists, $I_2 = o(\|x\|)$ as $\|x\| \to 0$ and *a fortiori* $I_2 = o(\|x\| + \|y\|) = o(\|(x, y)\|)$ as $\|(x, y)\| \to 0$. To estimate I_1, let $\epsilon > 0$ and choose $\delta > 0$ such that

$$\|\partial_y F[(x, y)] - \partial_y F[(x_0, y_0)]\| \leq \epsilon \text{ if } \|(x - x_0, y - y_0)\| \leq \delta.$$

For any x, y with $\|(x, y)\| < \delta$ let

$$u(t) = F(x_0 + x, y_0 + ty) - t\partial_y F[(x_0, y_0)]y.$$

Then, by the Chain Rule,

$$\|du[t]\| = \|(\partial_y F[(x_0 + x, y_0 + ty)] - \partial_y F[(x_0, y_0)])y\|$$
$$\leq \|\partial_y F[(x_0 + x, y_0 + ty)] - \partial_y F[(x_0, y_0)]\| \|y\|$$
$$\leq \epsilon \|y\| \text{ since } \|(x, y)\| < \delta.$$

Hence, by Lemma 3.1.4,

$$I_1 = \|u(1) - u(0)\| \leq \epsilon \|y\| \leq \epsilon(\|x\| + \|y\|)$$

if $\|(x, y)\| < \delta$. Hence $I_1 = o(\|(x, y)\|)$ as $\|(x, y)\| \to 0$. Combining these estimates for I_1 and I_2 yields that $dF[(x_0, y_0)]$ exists and $dF[(x_0, y_0)](x, y) = \partial_x F[(x_0, y_0)]x + \partial_y F[(x_0, y_0)]y$. This completes the proof. $\qquad \square$

Suppose that F is Fréchet differentiable on U and $x \mapsto dF[x]$ is continuous from $U \subset X$ into the Banach space $\mathcal{L}(X, Y)$. Then we say that F is continuously Fréchet differentiable on U, or that F is of class C^1 on U. This is written $F \in C^1(U, Y)$.

As a consequence of the preceding results, a function defined on a product of Banach spaces is continuously differentiable if and only if each of its partial derivatives is continuously differentiable. Here are three examples where C^1 functions arise; the first will be familiar and the third shows that nothing can be taken for granted.

EXAMPLE 3.1.8   (Finite Dimensions) An important special case of this observation occurs when $f : \mathbb{R}^N \to \mathbb{R}^M$ is given by

$$f(x_1, \cdots, x_N) = \left(f^1(x_1, \cdots, x_N), \cdots, f^M(x_1, \cdots, x_N) \right).$$

Then f is continuously Fréchet differentiable if and only if each $\partial_{x_j} f^i$, $1 \leq i \leq M$, $1 \leq j \leq N$, is a continuous $\mathbb{R}$-valued function of $(x_1, \cdots, x_N)$.

When $U \subset \mathbb{F}^N$ is open and $f : \mathbb{F}^N \to \mathbb{F}^M$, the Fréchet derivative of f at a point is sometimes called the total derivative to distinguish it from partial derivatives. From advanced calculus courses we may recall that the existence of a total derivative implies the existence of all the partial derivatives of the component functions of f at the same point, but not *vice versa*. However if all the partial derivatives exist and are continuous at every point of U, then the total derivative also exists and $f \in C^1(U, \mathbb{F}^M)$. We have now observed that this theory remains valid in infinite dimensions with no extra difficulties.   □

EXAMPLE 3.1.9   (Nemytskii Operators on C) In Examples 2.1.2 (b) we defined the Banach space $C([0,1], \mathbb{F}^N)$. Suppose that $f : \mathbb{F}^N \to \mathbb{F}^M$ is continuously differentiable. Then a Nemytskii operator $F : C([0,1], \mathbb{F}^N) \to C([0,1], \mathbb{F}^M)$ can be defined by composition:

$$F(u)(t) = f(u(t)), \quad t \in [0,1], \quad u \in C([0,1], \mathbb{F}^N),$$

and it is not difficult to see that the nonlinear operator F is of class C^1 in this setting.   □

EXAMPLE 3.1.10   (Nemytskii Operators on L_p) Let $L_p[0,1]$, $1 \leq p < \infty$, denote the Banach space of pth power Lebesgue integrable real-valued 'functions' $u : [0,1] \to \mathbb{F}$ with the norm

$$\|u\|_p = \left(\int_0^1 |u(s)|^p ds \right)^{1/p}.$$

Suppose that $g : \mathbb{R} \to \mathbb{R}$, $g(0) = 0$ and let

$$G(u)(x) = g(u(x)), \quad x \in [0,1], \quad u \in L_p[0,1].$$

Now suppose that G maps $L_p[0,1]$ into $L_p[0,1]$ and is Fréchet differentiable at $0 \in L_p[0,1]$. Then $g(t) = bt$ for some $b \in \mathbb{R}$. To see this, suppose that $g(s)/s \neq g(t)/t$

for some non-zero s, $t \in \mathbb{R}$. For $\delta \in [0, 1)$, let $u_\delta = t\chi_{[0,\delta]}$ and $v_\delta = s\chi_{[0,\delta]}$, where $\chi_{[a,b]}$ is the function with value 1 on $[a, b]$ and zero otherwise. Then $v_\delta = su_\delta/t$ and clearly $\|u_\delta\|_p$ and $\|v_\delta\|_p$ tend to 0 as $\delta \to 0$. Now suppose that $dG[0] = L \in \mathcal{L}(L_p[0, 1], L_p[0, 1])$ exists. Then, since $G(0) = 0$, $\|G(u_\delta) - Lu_\delta\|_p/\|u_\delta\|_p \to 0$ and

$$\frac{\|(t/s)G(v_\delta) - Lu_\delta\|_p}{\|u_\delta\|_p} = \frac{\|G(v_\delta) - Lv_\delta\|_p}{\|v_\delta\|_p} \to 0$$

as $\delta \to 0$. Thus

$$\frac{|(t/s)g(s) - g(t)|}{|t|} = \frac{\|(t/s)G(v_\delta) - G(u_\delta)\|_p}{\|u_\delta\|_p} \to 0 \text{ as } \delta \to 0.$$

This contradiction shows that, in the setting of pth power Lebesgue integrable functions, Fréchet differentiability at 0 of a Nemytskii operator implies that the operator in question is affine (linear + constant). $\square$

The notion of a compact linear operator in §2.7 has a natural analogue for nonlinear operators, although a certain amount of care should be exercised (see the remark following the definition).

DEFINITION 3.1.11 *A (nonlinear) function F from a subset U of a Banach space X into a Banach space Y is said to be compact if $\overline{F(W)}$ is compact in Y when $W \subset U$ is bounded in X.*

A compact linear operator from X to Y maps bounded sets to bounded sets and is therefore continuous. If F is nonlinear and compact, it need not be continuous. A continuous compact operator is sometimes called completely continuous. Many problems can be written as equations which involve nonlinear compact operators in Banach spaces. It is therefore useful to note that the linearizations of such equations involve compact linear operators.

LEMMA 3.1.12   (Differentiation of Compact Operators) *Suppose U is open in X, $F : U \to Y$ is compact, and $dF[x_0]$ exists at $x_0 \in U$. Then $dF[x_0] \in \mathcal{L}(X, Y)$ is compact.*

Proof. Let $\{x_n\} \subset X$ be bounded. Then the compactness of F and a diagonalization argument means that there is no loss of generality in supposing that $\{x_0 + (x_n/k)\} \subset U$ and $\{F(x_0 + (x_n/k))\}_{n \in \mathbb{N}}$ is convergent as $n \to \infty$ for each fixed $k \in \mathbb{N}$ with $k \geq K$ sufficiently large. We want to show that $\{dF[x_0](x_n)\}_{n \in \mathbb{N}}$ is Cauchy in Y. Let $M = \sup\{\|x_n\| : n \in \mathbb{N}\}$ and let $\epsilon > 0$ be given. Then there exists $\delta > 0$ such that

$$\|R(h)\| = \|F(x_0 + h) - F(x_0) - dF[x_0]h\| \leq \frac{\epsilon\|h\|}{2M} \text{ if } \|h\| < \delta.$$

For all n, m, $k \in \mathbb{N}$,

$$dF[x_0]x_n - dF[x_0]x_m = k\{dF[x_0](x_n/k) - dF[x_0](x_m/k)\}$$
$$= k\left\{R\left(\frac{x_m}{k}\right) - R\left(\frac{x_n}{k}\right) + F\left(x_0 + \frac{x_n}{k}\right) - F\left(x_0 + \frac{x_m}{k}\right)\right\}.$$

Let $\hat{k}$ (fixed) be such that $\hat{k} \geq K$ and $\hat{k} > M/\delta$. Then

$$\|dF[x_0]x_n - dF[x_0]x_m\|$$
$$\leq \hat{k}\left\{\|R(\frac{x_m}{\hat{k}})\| + \|R(\frac{x_n}{\hat{k}})\| + \|F(x_0 + \frac{x_n}{\hat{k}}) - F(x_0 + \frac{x_m}{\hat{k}})\|\right\}$$
$$\leq \hat{k}\left\{\frac{\epsilon}{2M\hat{k}}\{\|x_n\| + \|x_m\|\} + \|F(x_0 + \frac{x_n}{\hat{k}}) - F(x_0 + \frac{x_m}{\hat{k}})\|\right\}$$
$$\leq \epsilon + \hat{k}\|F(x_0 + (x_n/\hat{k})) - F(x_0 + (x_m/\hat{k}))\|.$$

Since $\{F(x_0 + (x_n/\hat{k}))\}_{n\in\mathbb{N}}$ is Cauchy, the sequence $\{dF[x_0]x_n\}_{n\in\mathbb{N}}$ is also Cauchy, and the result follows.   □

The converse is false: the derivative of a function at a point may be compact without the operator itself being compact. For example, let $F(u) = u^2$ for $u \in C[0,1]$. Then $dF[0] = 0 \in \mathcal{L}\big(C[0,1], C[0,1]\big)$, a compact operator. Now let $u_n(x) = 0$ for $x \in [n^{-1}, 1]$ and $u_n(x) = 1 - nx$ for $x \in [0, n^{-1}]$. Clearly $\{F(u_n)\}$ is not relatively compact in $C[0,1]$ and hence F is not a compact operator.

In finite-dimensional spaces, all continuous functions on closed sets are compact. While the compactness of F implies the compactness of $dF[x] \in \mathcal{L}(X, Y)$, it does not imply the compactness of $dF : X \to \mathcal{L}(X, Y)$, even in one dimension. Here is an example. Let $f(0) = 0$ and $f(x) = x^2 \sin(1/x^2)$ for $x \neq 0$. Since df is everywhere defined with $df[0] = 0$, but df is not bounded in a neighbourhood of 0, $df : \mathbb{R} \to \mathcal{L}(\mathbb{R}, \mathbb{R})$ is not compact.

3.2 HIGHER DERIVATIVES

Suppose that F is continuously Fréchet differentiable on U. If $dF : U \to \mathcal{L}(X, Y)$ is itself differentiable at $x_0 \in U$, we say that the second Fréchet derivative of F at $x_0 \in U$ exists. Note that

$$d\big(dF\big)[x_0] \in \mathcal{L}(X, \mathcal{L}(X, Y)),$$
$$d(dF)[x_0]x_1 \in \mathcal{L}(X, Y),$$
$$d(dF)[x_0](x_1)(x_2) \in Y, \text{ for all } x_1, x_2 \in X,$$

and the mappings $d(dF)[x_0](\cdot)(x_2)$ and $x_2 \mapsto d(dF)[x_0](x_1)(\cdot)$ are elements of $\mathcal{L}(X, Y)$.

THEOREM 3.2.1 *Suppose that $F : U \to Y$ and the second Fréchet derivative of F exists at $x_0 \in U$ (open in X). Then*

$$d(dF)[x_0](x_1)(x_2) = d(dF)[x_0](x_2)(x_1), \quad (x_1, x_2) \in X^2.$$

Proof. For x_1, x_2 sufficiently close to 0 in X let

$$\Phi(x_1, x_2) = F(x_0 + x_1 + x_2) - F(x_0 + x_1) - F(x_0 + x_2) + F(x_0).$$

Clearly $\Phi(x_1, x_2) = \Phi(x_2, x_1)$. To prove the required result it suffices to show that

$$\|\Phi(x_2, x_1) - d(dF)[x_0](x_1)(x_2)\| = o(\|x_1\|^2 + \|x_2\|^2)$$

as $\|x_1\| + \|x_2\| \to 0$, because, for any $t \in \mathbb{R}$ and $x_1, x_2 \in X$ (fixed) it then follows that

$$\begin{aligned}
t^2 \| & d(dF)[x_0](x_1)(x_2) - d(dF)[x_0](x_2)(x_1)\| \\
&= \|d(dF)[x_0](tx_1)(tx_2) - d(dF)[x_0](tx_2)(tx_1)\| \\
&= \|\big(\Phi(tx_1, tx_2) - d(dF)[x_0](tx_2)(tx_1)\big) \\
&\quad - \big(\Phi(tx_2, tx_1) - d(dF)[x_0](tx_1)(tx_2)\big)\| \\
&= o(\|tx_1\|^2 + \|tx_2\|^2) = o(t^2) \text{ as } t \to 0.
\end{aligned}$$

Dividing through by t^2 and letting $t \to 0$ yields the required result. Now

$$\begin{aligned}
\|\Phi(&x_2, x_1) - d(dF)[x_0](x_1)(x_2)\| \\
&\leq \|\Phi(x_2, x_1) - dF[x_0 + x_1]x_2 + dF[x_0]x_2\| \\
&\quad + \|dF[x_0 + x_1]x_2 - dF[x_0]x_2 - d(dF)[x_0](x_1)(x_2)\| \\
&= I_1(x_1, x_2) + I_2(x_1, x_2), \text{ say,}
\end{aligned}$$

and

$$\begin{aligned}
I_2(x_1, x_2) &\leq \|dF[x_0 + x_1] - dF[x_0] - d(dF)[x_0]x_1\| \|x_2\| \\
&= \|x_2\|(o(\|x_1\|)) \text{ as } \|x_1\| \to 0,
\end{aligned}$$

since $d(dF)[x_0]$ is the derivative of $dF : X \to \mathcal{L}(X, Y)$ at $x_0 \in X$. Hence for $\epsilon > 0$ there exists $\delta > 0$ such that

$$I_2(x_1, x_2) \leq \epsilon \|x_1\| \|x_2\| \text{ if } \|x_1\| \leq \delta,$$

whence

$$I_2(x_1, x_2) \leq \tfrac{1}{2}\epsilon(\|x_1\|^2 + \|x_2\|^2) \text{ if } \|x_1\| + \|x_2\| \leq \delta.$$

This shows that

$$I_2(x_1, x_2) = o(\|x_1\|^2 + \|x_2\|^2) \text{ as } \|x_1\| + \|x_2\| \to 0.$$

For $t \in [0, 1]$ let

$$u(t) = F(x_0 + x_1 + tx_2) - F(x_0 + tx_2) - t(dF[x_0 + x_1]x_2 - dF[x_0]x_2).$$

Then

$$I_1(x_1, x_2) = \|u(1) - u(0)\| \leq \sup\{\|du[t]\| : 0 \leq t \leq 1\},$$

by Lemma 3.1.4. Now $\|du[t]\|$ is bounded by

$$\|x_2\|\|dF[x_0 + x_1 + tx_2] - dF[x_0 + tx_2] - dF[x_0 + x_1] + dF[x_0]\|$$
$$= \|x_2\|\|\{dF[x_0 + x_1 + tx_2] - dF[x_0] - d(dF)[x_0](x_1 + tx_2)\}$$
$$- \{dF[x_0 + tx_2] - dF[x_0] - d(dF)[x_0](tx_2)\}$$
$$- \{dF[x_0 + x_1] - dF[x_0] - d(dF)[x_0]x_1\}\|$$
$$= \|x_2\|\{o(\|x_1 + tx_2\|) + o(\|tx_2\|) + o(\|x_1\|)\}$$

as $\|x_1\| + \|x_2\| \to 0$. Hence given $\epsilon > 0$ there exists $\delta > 0$ such that if $\|x_1\| + \|x_2\| \le \delta$, then

$$I_1(x_1, x_2) \le \epsilon\|x_2\|(\|x_1 + tx_2\| + \|tx_2\| + \|x_1\|)$$
$$\le 2\epsilon\|x_2\|(\|x_1\| + \|x_2\|) \le 3\epsilon(\|x_1\|^2 + \|x_2\|^2).$$

Therefore $I_1(x_1, x_2) = o(\|x_1\|^2 + \|x_2\|^2)$ as $\|x_1\| + \|x_2\| \to 0$. Combining the estimates for I_1 and I_2 yields the estimate for $\|\Phi(x_2, x_1) - d(dF)[x_0](x_1)(x_2)\|$ needed to complete the proof.   □

If the second Fréchet derivative exists at every point of U we say that F is twice differentiable on U, and if $d(dF)$ is continuous from U to $\mathcal{L}(X, \mathcal{L}(X, Y))$ we say that F is twice continuously differentiable on U. We then write $F \in C^2(U, Y)$ and say that F is of class C^2 on U. It is usual to write $d(dF)[x_0](x_1)(x_2)$ as $d^2 F[x_0](x_1, x_2)$. Remember that the order of x_1 and x_2 does not matter.

DEFINITION 3.2.2 *An operator $b : X \times X \to Y$ with the property that $b(x, \cdot)$ and $b(\cdot, x)$ are linear on X for each fixed $x \in X$ is said to be $\mathbb{F}$-bilinear. If in addition, $b(x_1, x_2) = b(x_2, x_1)$ for all $(x_1, x_2) \in X \times X$, b is called a symmetric $\mathbb{F}$-bilinear operator from X into Y. When $Y = \mathbb{F}$, b is called a bilinear form. When the field $\mathbb{F}$ is given by the context, we can omit the prefix $\mathbb{F}$ and refer to bilinear forms and operators.*

When $d^2 F[x]$ exists it is a symmetric bilinear operator from $X \times X$ to Y. It is clear that the nth Fréchet derivative of F can be defined by induction. For $x_0 \in U$, $x_1, \cdots, x_n \in X$ and $1 \le k \le n$, let $d^k F[x_0](x_1, \cdots, x_k)$ be defined recursively as follows.

$$d^1 F[x_0](x_1) = dF[x_0]x_1,$$

$$d^2 F[x_0](x_1, x_2) = d(dF)[x_0](x_1)(x_2),$$

$$d^k F[x_0](x_1, \cdots, x_k) = \underbrace{d(\cdots d(dF) \cdots)}_{k-1 \text{ parentheses}}[x_0](x_1)(x_2) \cdots (x_k).$$

PROPOSITION 3.2.3 *Suppose F maps a neighbourhood U of x_0 in X into Y and that the nth Fréchet derivative of F at x_0 exists. Then*

$$x \mapsto d^n F[x_0](x_1, \cdots, x_{k-1}, x, x_{k+1}, x_n), \quad x \in X,$$

is linear for each $k \in \{1, \cdots, n\}$, and

$$d^n F[x_0](x_1, \cdots, x_n) = d^n F[x_0](x_{\pi(1)}, \cdots, x_{\pi(n)})$$

for all $\pi \in S_n$ (the symmetric group).

Proof. By definition, the nth derivative of F at x_0,

$$d^n F[x_0] \in \mathcal{L}(X, \mathcal{L}(X, \mathcal{L}(X, \cdots, \mathcal{L}(X, Y) \cdots)),$$

($\mathcal{L}$ repeated n times) is identified with $d^n F[x_0](x_1, x_2, \cdots, x_n)$. It is clear from the definition that $d^n F[x_0]$ is linear in each of the variables $(x_1, \cdots, x_n)$ separately. We need only show its symmetry.

Theorem 3.2.1 shows the result for $n = 2$. Suppose $n \geq 3$ and $(x_3, \cdots, x_{n+1}) \in X^{n-1}$ and define $g : U \to Y$ by

$$g(x) = d^{n-1} F[x](x_3, \cdots, x_{n+1}).$$

By Theorem 3.2.1, for any $(x_1, x_2) \in X \times X$,

$$d^{n+1} F[x_0](x_1, x_2, \cdots, x_{n+1}) = d^2 g[x_0](x_1, x_2)$$
$$= d^2 g[x_0](x_2, x_1) = d^{n+1} F[x_0](x_2, x_1, \cdots, x_{n+1}). \quad (3.1)$$

We proceed by induction. Suppose that $d^{n+1} F[x_0]$ exists and that for all $x \in U$ and $(x_1, \cdots, x_n) \in X^n$,

$$d^n F[x](x_1, \cdots, x_n) = d^n F[x](x_{\pi(1)}, \cdots, x_{\pi(n)}),$$

where $\pi \in S_n$, the symmetric group. A differentiation with respect to x at x_0 now gives, for $x \in X$,

$$d^{n+1} F[x_0](x, x_1, \cdots, x_n) = d^{n+1} F[x_0](x, x_{\pi(1)}, \cdots, x_{\pi(n)}). \quad (3.2)$$

From (3.1) and (3.2) it follows that $d^{n+1} F[x_0]$ is symmetric with respect to interchanging any pair of its components. This completes the proof. $\square$

REMARK 3.2.4 (Higher Mixed Derivatives) Suppose that the second Fréchet derivative of $F : U \subset X \times Y \to Z$ exists at (x_0, y_0). Then $\partial_x F : U \to \mathcal{L}(X, Z)$ is well defined and has a partial derivative with respect to y

$$\partial_y (\partial_x F)[(x_0, y_0)] \in \mathcal{L}(Y, \mathcal{L}(X, Z)).$$

A similar statement holds when x and y are interchanged. It is clear from the definitions that for $x \in X$, $y \in Y$,

$$\left(\partial_y (\partial_x F)[(x_0, y_0)](y) \right)(x) = d^2 F[(x_0, y_0)]((x, 0), (0, y))$$
$$= d^2 F[(x_0, y_0)]((0, y), (x, 0)) = \left(\partial_x (\partial_y F)[(x_0, y_0)](x) \right)(y).$$

When F has a second Fréchet derivative, $\partial_y (\partial_x F) = \partial_x (\partial_y F)$ in this sense and we denote it by $\partial^2_{x,y} F[(x_0, y_0)] = \partial^2_{y,x} F[(x_0, y_0)]$. The second partial derivative with respect to x is denoted by $\partial^2_{x^2} F[(x_0, y_0)]$. $\square$

3.3 TAYLOR'S THEOREM

Here is the extension to a Banach space setting of the classical theorem on the difference between a function and its nth order Taylor polynomial, familiar from real-variable theory. For $x \in X$ and $k \in \mathbb{N}_0$ define

$$d^k F[x_0]x^k = d^k F[x_0] \underbrace{(x, \cdots, x)}_{k \text{ times}}, \quad x_0 \in U,$$

with the convention that

$$d^0 F[x_0]x^0 = F(x_0), \quad x \in X, \; x_0 \in U.$$

THEOREM 3.3.1 **(Taylor's Theorem)** *Suppose X, Y Banach spaces, $U \subset X$ open and convex, and $F \in C^{n+1}(U, Y)$, $n \in \mathbb{N}_0$. Let x, $x_0 \in U$. Then*

$$F(x) - \sum_{k=0}^{n} \frac{1}{k!} d^k F[x_0](x - x_0)^k = R_n(x, x_0)$$

where

$$\|R_n(x, x_0)\| \leq \frac{\|x - x_0\|^{n+1}}{(n+1)!} \sup_{0 \leq t \leq 1} \|d^{n+1} F[(1 - t)x_0 + tx]\|.$$

Proof. Suppose that $f : [0, 1] \to \mathbb{R}$ has $n + 1$ derivatives which are continuous on $[0, 1]$. Then it is elementary to prove, using induction and the fundamental theorem of calculus, that, for $t \in [0, 1]$,

$$f(t) - \sum_{k=0}^{n} \frac{f^{(k)}(0)t^k}{k!}$$

$$= \int_0^t \int_0^{x_1} \int_0^{x_2} \cdots \int_0^{x_n} f^{(n+1)}(s) ds dx_n \cdots dx_2 dx_1,$$

and therefore

$$\left| f(t) - \sum_{k=0}^{n} \frac{f^{(k)}(0)t^k}{k!} \right| \leq \frac{t^{n+1}}{(n+1)!} \sup\{|f^{n+1}(s)| : s \in [0, 1]\}.$$

Now for x, x_0 in the statement let $y^* \in Y^*$ be such that $\|y^*\| = 1$ and

$$y^* \left(F(x) - \sum_{k=0}^{n} \frac{1}{k!} d^k F[x_0](x - x_0)^k \right)$$

$$= \left\| F(x) - \sum_{k=0}^{n} \frac{1}{k!} d^k F[x_0](x - x_0)^k \right\|,$$

and, for $t \in [0, 1]$, let

$$f(t) = \Re \left\{ y^* \left(F \left((tx + (1 - t)x_0) \right) \right) \right\}.$$

Then in the statement of the theorem,

$$\| R_n(x - x_0) \| = y^* \left(F(x) - \sum_{k=0}^{n} \frac{1}{k!} d^k F[x_0](x - x_0)^k \right)$$

$$= f(1) - \sum_{k=0}^{n} \frac{f^{(k)}(0)1^k}{k!} \leq \frac{1}{(n+1)!} \sup\{ |f^{n+1}(s)| : s \in [0, 1] \}$$

$$\leq \frac{\|x - x_0\|^{n+1}}{(n+1)!} \sup_{0 \leq t \leq 1} \| d^{n+1} F[(1 - t)x_0 + tx] \|,$$

because

$$f^{(n+1)}(t) = \Re \left\{ y^* \left(d^{n+1} F[(1 - t)x + tx_0](x - x_0)^{n+1} \right) \right\}.$$

This completes the proof. $\qquad\qquad\qquad\qquad\qquad\qquad\qquad\qquad\square$

DEFINITION 3.3.2 *Suppose that X and Y are Banach spaces and $U \subset X$ is open. If $F : U \to Y$ has Fréchet derivatives of all orders up to n at $x_0 \in U$, then the nth order Taylor polynomial of F at x_0 is*

$$T_n[F]_{x_0}(x) = \sum_{k=0}^{n} \frac{1}{k!} d^k F[x_0](x - x_0)^k.$$

When it is defined the infinite series

$$\sum_{k=0}^{\infty} \frac{1}{k!} d^k F[x_0](x - x_0)^k,$$

whether it converges or not, is called the Taylor series of F at x_0.

When it does converge, there is no *a priori* reason for its limit to be $F(x)$; in general it is not.

3.4 GRADIENT OPERATORS

Before the general case, we consider the real Banach space $\mathbb{R}^n$ where it is common to speak of gradient vectors and Jacobian matrices in the context of differentiation.

Let $\langle \cdot, \cdot \rangle$ denote an inner product on $\mathbb{R}^n$. If $g : \mathbb{R}^n \to \mathbb{R}$ is Fréchet differentiable at x_0 its derivative is an element of $\mathcal{L}(\mathbb{R}^n, \mathbb{R}) = \mathbb{R}^{n*}$. Hence, by the Riesz representation theorem, there exists a unique y_0 in $\mathbb{R}^n$ such that

$$dg[x_0]x = \langle y_0, x \rangle \text{ for all } x \in \mathbb{R}^n.$$

We say that $y_0 \in \mathbb{R}^n$ is the gradient of g at x_0 with respect to the inner product $\langle \cdot, \cdot \rangle$. However there are many inner products on $\mathbb{R}^n$. For example, if $\{e_1, \cdots, e_n\}$ is a basis for $\mathbb{R}^n$ then

$$\langle x, y \rangle = \sum_{i=1}^{n} \alpha_i \beta_i \text{ where } x = \sum_{i=1}^{n} \alpha_i e_i \text{ and } y = \sum_{i=1}^{n} \beta_i e_i$$

defines a different inner product for each choice of basis. Clearly the gradient depends on the choice of inner product. With the above inner product,

$$\langle y_0, x \rangle = dg[x_0]x = dg[x_0]\left(\sum_{i=1}^{n} \alpha_i e_i \right) = \sum_{i=1}^{n} \alpha_i dg[x_0]e_i$$

$$= \left\langle \sum_{i=1}^{n} \alpha_i e_i, \sum_{i=1}^{n} (dg[x_0]e_i)e_i \right\rangle,$$

and hence the gradient of g at x_0 is given by

$$\sum_{i=1}^{n} (dg[x_0]e_i)e_i \in \mathbb{R}^n.$$

If the standard basis and standard inner product on $\mathbb{R}^n$ have been chosen, then the directional derivative $dg[x_0]e_i$ is given by the classical partial derivative $\partial g/\partial x_i \big|_{x_0}$ and so the gradient of g is given by the familiar formula

$$\nabla g(x_0) = (\partial g/\partial x_1, \cdots, \partial g/\partial x_n)\big|_{x_0},$$

where it is understood that $g(x)$ is given as a function of the components of x with respect to the standard basis.

Now we consider the second derivative $d^2 g[x_0]$. This is a symmetric bilinear form on $\mathbb{R}^n$. Therefore, for each $x \in \mathbb{R}^n$, the Riesz representation theorem implies the existence of a unique point y_x such that

$$d^2 g[x_0](x, y) = \langle y_x, y \rangle, \text{ for all } y \in \mathbb{R}^n$$

and y_x depends linearly on x. Hence there exists a linear transformation L on $\mathbb{R}^n$ such that $Lx = y_x$. Since $d^2 g[x_0]$ is symmetric and $\mathbb{F} = \mathbb{R}$ in this example,

$$\langle Lx, y \rangle = d^2 g[x_0](x, y) = d^2 g[x_0](y, x) = \langle Ly, x \rangle = \langle x, Ly \rangle.$$

Hence L, the Jacobian transformation of g at x_0, is a symmetric operator. The matrix which represents L with respect to the basis used in the definition of the inner product is called the Jacobian matrix. As before, when the standard basis and inner product are chosen, we find a familiar formula for the components of the Jacobian matrix

$$L_{i,j} = (\partial^2 g(x_0)/\partial x_i \partial x_j)\big|_{x_0}, \quad i, j = 1, \cdots, n.$$

Now suppose that X is a Hilbert space over $\mathbb{F}$ with inner product $\langle \cdot, \cdot \rangle$ and suppose that $U \subset X$ is open. If $g : U \to \mathbb{F}$ is Fréchet differentiable at $x_0 \in U$ then $dg[x_0] \in X^*$ and, by the Riesz representation theorem, there exists $\nabla g(x_0) \in X$ such that

$$dg[x_0]y = \langle \nabla g(x_0), y \rangle \text{ for all } y \in X.$$

The element $\nabla g(x_0)$ is called the gradient of g at x_0 (where the rôle of X is understood). If $G : U \to X$ coincides with ∇g on U we say that G is a gradient operator. More generally we have the following definition.

DEFINITION 3.4.1 *Suppose that Y is a Banach space over $\mathbb{F}$ which is continuously embedded in a Hilbert space $(X, \langle \cdot, \cdot \rangle)$, let $U \subset Y$ be open and let $g : U \to \mathbb{F}$ be Fréchet differentiable at $y_0 \in U$. Then $dg[y_0] \in \mathcal{L}(Y, \mathbb{F}) = Y^*$. Suppose that there exists $x_g \in X$ such that*

$$dg[y_0]y = \langle x_g, y \rangle, \text{ for all } y \in Y,$$

then we say that x_g is the gradient in X of g at y_0 and write $x_g = \nabla_X g(y_0)$. Therefore, when it exists,

$$dg[y_0]y = \langle \nabla_X g(y_0), y \rangle \text{ for all } y \in Y.$$

(When X is clear from the context, we use ∇ instead of ∇_X for the gradient.)

Now suppose that $g : U \to \mathbb{F}$ has a gradient in X, and that $\nabla_X g : U \to X$ has a derivative $d(\nabla_X g)[y_0]$, which we denote by $D^2 g[y_0] \in \mathcal{L}(Y, X)$, at every point $y_0 \in U$. Since, for $y \in Y$, $D^2 g[y_0]y \in X$ and Y is continuously embedded in X, we can define $\ell_{y_0} y \in Y^*$ by

$$\ell_{y_0} y(z) = \langle D^2 g[y_0]y, z \rangle \text{ for all } z \in Y$$

where

$$|\ell_{y_0} y(z)| \leq \|D^2 g[y_0]\|_{\mathcal{L}(Y,X)} \|y\|_Y \|z\|_Y \text{ for all } z \in Y.$$

Since Y is continuously embedded in X, for $h \in Y$ such that $y_0 + h \in U$,

$$\left(\frac{dg[y_0 + h] - dg[y_0] - \ell_{y_0} h}{\|h\|} \right) z$$

$$= \left\langle \frac{\nabla_X g(y_0 + h) - \nabla_X g(y_0) - D^2 g[y_0]h}{\|h\|}, z \right\rangle \to 0 \text{ as } \|h\|_Y \to 0,$$

uniformly for $z \in Y$ with $\|z\|_Y \leq 1$. Therefore

$$d^2 g[y_0](y, x) = \ell_{y_0} y(x) = \langle D^2 g[y_0]y, x \rangle.$$

DEFINITION 3.4.2 *If $G : \mathbb{R} \times Y \to X$ has the property that*

$$G(\lambda, \cdot) = \nabla_X g(\lambda, \cdot)$$

where $g : \mathbb{R} \times Y \to \mathbb{R}$ is C^2, then the equation $G(\lambda, y) = 0$ is said to be the Euler-Lagrange equation of the functional g. Equivalently we say that the equation has gradient structure.

In the theory of nonlinear equations, those with gradient structure have important special properties. We return to this in §11.3.

3.5 INVERSE AND IMPLICIT FUNCTION THEOREMS

The inverse function theorem says that if the Fréchet derivative of a nonlinear operator F at x_0 is invertible, then the nonlinear operator itself is a bijection from an open neighbourhood of x_0 onto an open neighbourhood of $F(x_0)$. Before giving its precise statement we prove a technical result.

LEMMA 3.5.1 *Let X and Y be Banach spaces, $U \subset X$ a convex open set, and let $F : U \to Y$ be Fréchet differentiable at each point of U. Suppose that there exists $A \in \mathcal{L}(X, Y)$ such that $\|dF[x] - A\| \leq M$ for all $x \in U$. Then for all $x_1, x_2 \in U$,*

$$\|F(x_1) - F(x_2) - dF[x_2](x_1 - x_2)\| \leq 2M\|x_1 - x_2\|.$$

Proof. Let $x_2 \in U$ be arbitrary, but fixed. We apply the Lemma 3.1.4 to G defined on U by $G(x) = F(x) - dF[x_2]x$, $x \in U$. Then

$$\|dG[x]\| = \|dF[x] - dF[x_2]\| \leq \|dF[x] - A\| + \|A - dF[x_2]\| \leq 2M.$$

Therefore, by Lemma 3.1.4,

$$\|F(x_1) - F(x_2) - dF[x_2](x_1 - x_2)\|$$
$$= \|G(x_1) - G(x_2)\| \leq 2M\|x_1 - x_2\|.$$

This proves the lemma. □

THEOREM 3.5.2 (Inverse Function Theorem) *Let $x_0 \in U$, an open subset of a Banach space X, and let $F \in C^1(U, Z)$ where Z is also a Banach space. Suppose that $dF[x_0] \in \mathcal{L}(X, Z)$ is a homeomorphism.*

Then there exists a connected open set $U_0 \subset U$ with $x_0 \in U_0$ and an open ball $W \subset Z$ with $F(x_0) \in W$, such that $F : U_0 \to W$ is a bijection and $F^{-1} \in C^1(W, X)$.

If, in addition, $F \in C^k(U, Z)$, $k \in \mathbb{N}$, then $F^{-1} \in C^k(W, X)$.

REMARK 3.5.3 We say that $F\big|_{U_0}$ is a diffeomorphism onto W. □

Proof. By considering the mapping $x \mapsto dF[x_0]^{-1}(F(x_0 + x) - F(x_0))$ in place of F, there is no loss of generality in supposing that $X = Z$, that $x_0 = F(x_0) = 0 \in X$, and that $dF[0] = I \in \mathcal{L}(X, X)$, the identity operator. From this special case the general result follows.

In the new setting, let $r \in (0, 1)$ be chosen so that if $\|x\| \leq r$ then $x \in U$ and $\|dF(x) - I\| < 1/4$. There is no further loss of generality in supposing that $U = \{x : \|x\| < r\}$. We will now show that, if $y \in X$ is sufficiently close to zero, a sequence $\{x_n\} \subset U$ can be defined inductively by the recipe

$$x_0 = 0, \quad x_{n+1} = y + x_n - F(x_n),$$

and the sequence so defined converges to a solution x of the equation $F(x) = y$. Now, provided that $x_n \in U$ for all $n \in \mathbb{N}$ with $n \leq k$, the definition of x_n and Lemma 3.1.4 give that $\|x_1\| = \|y\|$ and,

$$
\begin{aligned}
\|x_{n+1} - x_n\| &= \|(F(x_{n-1}) - x_{n-1}) - (F(x_n) - x_n)\| \\
&\leq \|x_n - x_{n-1}\| \sup_{0 \leq t \leq 1} \|dF[x_{n-1} + t(x_n - x_{n-1})] - I\| \\
&\leq \frac{1}{4}\|x_n - x_{n-1}\|,
\end{aligned}
$$

for all $1 \leq n \leq k$.

Now choose y with $\|y\| < 3r/4$. It can be seen, by induction, that for all $n \geq 0$,

$$\|x_{n+1} - x_n\| \leq 4^{-n}\|y\| \quad \text{and} \quad \|x_n\| \leq 4\|y\|/3 < r.$$

Moreover, for all $m > n$

$$\|x_n - x_m\| \leq \sum_{k=n}^{m-1} \|x_{k+1} - x_k\| \leq 4^{-(n-1)}\|y\|$$

which converges to 0 as $n \to \infty$. Therefore $\{x_n\}$ is a Cauchy sequence which converges in the Banach space X, to x, say, and $\|x\| \leq 4\|y\|/3 < r$. By continuity, $x = y + x - F(x)$, whence $F(x) = y$. Let

$$W = \{y \in X : \|y\| < 3r/4\}, \quad U_0 = \{x : \|x\| < r \text{ and } F(x) \in W\}.$$

Then U_0 is open, W is an open ball, and $F : U_0 \to W$ is a bijection. Suppose that $y_1, y_2 \in W$, $x_1, x_2 \in U_0$, $F(x_1) = y_1$ and $F(x_2) = y_2$. Then

$$
\begin{aligned}
\|y_2 - y_1\| &= \|F(x_2) - F(x_1)\| \\
&= \|(x_2 - x_1) + (dF[x_2] - I)(x_2 - x_1) \quad\quad\quad (3.3) \\
&\quad + (F(x_2) - F(x_1) - dF[x_2](x_2 - x_1))\| \\
&\geq \|(x_2 - x_1)\| - \|(dF[x_2] - I)(x_2 - x_1)\| \\
&\quad - \|(F(x_2) - F(x_1) - dF[x_2](x_2 - x_1)\| \geq \frac{1}{4}\|x_2 - x_1\|, \quad (3.4)
\end{aligned}
$$

because of the choice of r and Lemma 3.5.1. This shows that F^{-1} is Lipschitz continuous on W. Hence $U_0 = F^{-1}(W) \subset U$ is connected.

Now we show that F^{-1} is differentiable at each point $y \in W$ and that the derivative $d(F^{-1})[y]$ depends continuously on $y \in W$. Let $y \in W$ and let $k \in Y$ be such that $y + k \in W$. Suppose that $F(x) = y$ and $F(x + h) = y + k$, where x, $x + h \in U_0$. Since $\|x\| < r$ it follows that $dF[x] = dF[F^{-1}(y)]$ is a homeomorphism (Lemma 2.5.1). Therefore

$$\frac{\left\| F^{-1}(y + k) - F^{-1}(y) - \left(dF[F^{-1}(y)] \right)^{-1} k \right\|}{\|k\|}$$

$$= \frac{\left\| (x + h) - x - \left(dF[x] \right)^{-1} (F(x + h) - F(x)) \right\|}{\|k\|}$$

$$= \frac{\left\| \left(dF[x] \right)^{-1} \{ F(x + h) - F(x) - dF[x]h \} \right\|}{\|h\|} \cdot \frac{\|h\|}{\|k\|}$$

$$\leq \left\| \left(dF[x] \right)^{-1} \right\| \frac{\| F(x + h) - F(x) - dF[x]h \|}{\|h\|} \cdot \frac{\|h\|}{\|k\|} \to 0$$

as $\|k\| \to 0$ by the definition of $dF[x]$ and the fact that $4\|k\| \geq \|h\|$, which follows from (3.4). This shows that, for all $y \in W$, F^{-1} is differentiable at y with

$$d(F^{-1})[y] = \left(dF[F^{-1}(y)] \right)^{-1}.$$

The continuity of $d(F^{-1})$ is immediate from Lemma 2.5.1.

Finally, from the formula for $d(F^{-1})[y]$ and the chain rule it follows that if $F \in C^k(U, X)$ then $F^{-1} \in C^k(W, X)$. $\qquad\square$

The next result is a corollary of the inverse function theorem.

THEOREM 3.5.4   (Implicit Function Theorem) *Let X, Y and Z be Banach spaces and suppose that $U \subset X \times Y$ is open. Let $(x_0, y_0) \in U$. Suppose also that, for some $k \in \mathbb{N}$, $F \in C^k(U, Z)$, that $F(x_0, y_0) = z_0$, and that the partial derivative $\partial_x F[(x_0, y_0)] \in \mathcal{L}(X, Z)$ is a homeomorphism.*

Then there exists an open ball $V \subset Y$ with centre $y_0 \in Y$, a connected open set $W \subset U$ and a mapping $\phi \in C^k(V, X)$ such that

$$(x_0, y_0) \in W \text{ and } F^{-1}(z_0) \cap W = \{ (\phi(y), y) : y \in V \}.$$

Proof. Define a new function $G \in C^k(U, Z \times Y)$ by

$$G(x, y) = \left(F(x, y), y \right).$$

Clearly $G(x_0, y_0) = (z_0, y_0)$ and

$$dG[(x_0, y_0)](x, y) = \left(\partial_x F[(x_0, y_0)]x + \partial_y F[(x_0, y_0)]y, \; y \right)$$

for $(x, y) \in X \times Y$. Therefore $dG[(x_0, y_0)]$ has a bounded inverse $dG[(x_0, y_0)]^{-1} :$ $Z \times Y \to X \times Y$ given by

$$(dG[(x_0, y_0)])^{-1}(z, y) = \left((\partial_x F[(x_0, y_0)])^{-1}(z - \partial_y F[(x_0, y_0)]y), \; y \right)$$

for $(x, y) \in X \times Y$, and one may apply the inverse function theorem 3.5.2 to G to obtain an open connected set $W \subset U$ and an open ball $R \subset Z \times Y$ such that $(x_0, y_0) \in W$, (z_0, y_0) is the centre of R, and $G : W \to R$ is a diffeomorphism of class $C^k(W, R)$ for the same k as in the statement of the theorem.

It suffices now to put $V = \{y : (z_0, y) \in R\}$ and to say that $\phi(y) = x$ for $y \in V$ if and only if $G^{-1}(z_0, y) = (x, y) \in W$. Then $(x, y) \in W$ and $F(x, y) = z_0$ if and only if $(z_0, y) \in R$ and $G^{-1}(z_0, y) = (x, y) \in W$. Now G^{-1} is of class C^k on R implies that $y \mapsto G^{-1}(z_0, y)$ is of class C^k on V. Let P denote the projection of $X \times Y$ onto $X \times \{0\}$. Then, since $(\phi(y), 0) = P(G^{-1}(z_0, y))$, it follows that ϕ is of class C^k on V. This completes the proof. $\qquad\square$

DEFINITION 3.5.5   *A set $M \subset \mathbb{F}^n$ is called an m-dimensional manifold of class C^k, or a C^k-manifold, $k \geq 1$, if, for all points $x \in M$, there is an open neighbourhood U_x of $0 \in \mathbb{F}^m$ and a function $f : U_x \to M$ such that $f(0) = x$, $df[0]$ is a finite-dimensional linear transformation of rank m and f maps open sets in U onto relatively open sets in M.*

REMARK 3.5.6   Suppose that a mapping $F : \mathbb{F}^n \times \mathbb{F}^m \to \mathbb{F}^n$ is of class C^k with $F(x_0, y_0) = z_0$ and that $\partial_x F[(x_0, y_0)]$ is a bijection on $\mathbb{F}^n$. Then the implicit function theorem 3.5.4 defines a C^k-manifold of dimension m by the equation $F(x, y) = z_0$ for (x, y) in a neighbourhood of (x_0, y_0). $\qquad\square$

3.6 PERTURBATION OF A SIMPLE EIGENVALUE

The following corollary of the implicit function theorem is often useful even in finite-dimensional linear algebra. The notation is that of Definition 2.7.8 and Lemma 2.7.9.

PROPOSITION 3.6.1   *Let $X \subset Y$ be Banach spaces and let the embedding operator $\iota \in \mathcal{L}(X, Y)$. Let $s \mapsto L(s)$ be a mapping of class C^k, $k \geq 1$, from $(-1, 1)$ into $\mathcal{L}(X, Y)$.*

Suppose that μ_0 is a simple eigenvalue of $L(0)$ with eigenvector $\xi_0 \in X$ where $\|\iota\xi_0\|_Y = 1$. Then there exists $\epsilon > 0$ and a C^k-curve

$$\{(\mu(s), \xi(s)) : s \in (-\epsilon, \epsilon)\} \subset \mathbb{R} \times X$$

such that $(\mu(0), \xi(0)) = (\mu_0, \xi_0)$,

$$L(s)\xi(s) = \mu(s)\iota\,\xi(s) \ \text{ and } \ \xi(s) = \xi_0 + \eta(s),$$

where $\iota\eta(s) \in \text{range}\,(L(0) - \mu_0\iota)$. Moreover, $\mu(s)$ is a simple eigenvalue of $L(s)$ and if $|s| < \epsilon$ and μ is an eigenvalue of $L(s)$ with $|\mu_0 - \mu| < \epsilon$ then $\mu = \mu(s)$.

Proof. Define a mapping $G : \mathbb{R} \times X \times (-1, 1) \to Y \times \mathbb{R}$ by

$$G(\mu, x, s) = \big(\mu\iota\,x - L(s)x, y^*(\iota x) - 1\big),$$

where $y^* \in Y^*$ is chosen using Corollary 2.3.1 (b) so that $y^*(\iota\xi_0) = 1$ and $y^*(\text{range}(\mu_0\,\iota - L(0))) = 0$. Then $G(\mu_0, \xi_0, 0) = (0,0)$ and

$$\partial_{(\mu,x)}G[(\mu_0,\xi_0,0)](\mu,x) = \left(\mu\iota\,\xi_0 + \mu_0\iota x - L(0)x, \, y^*(\iota x)\right)$$

for $(\mu, x) \in \mathbb{R} \times X$. Suppose that $\partial_{(\mu,x)}G[(\mu_0,\xi_0,0)](\mu,x) = (0,0)$. Then $\mu\iota\,\xi_0 \in \text{range}(\mu_0\iota - L(0))$ and, since μ_0 is a simple eigenvalue of $L(0)$, $\mu = 0$. Therefore $(L(0) - \mu_0\iota)x = 0$. Since μ_0 is simple, $x \in \text{span}\{\xi_0\}$ and, since $y^*(\iota x) = 0$ it follows that $x = 0$. Thus $\partial_{(\mu,x)}G[(\mu_0,\xi_0,0)]$ is injective. The surjectivity of $\partial_{(\mu,x)}G[(\mu_0,\xi_0,0)]$ follows immediately from the fact that, since μ_0 is a simple eigenvalue of $L(0)$,

$$Y = \text{range}(\mu_0\iota - L(0)) \oplus \text{span}\{\iota\xi_0\}.$$

Since $\partial_{(\mu,x)}G[(\mu_0,\xi_0,0)]$ is therefore a bijection, it is a homeomorphism by Corollary 2.4.3 and the implicit function theorem 3.5.4 gives the existence of a C^k curve $\{(\mu(s),\xi(s)) : s \in (-\epsilon,\epsilon)\} \subset \mathbb{R} \times X$ of solutions of $G(\mu, x, s) = (0,0)$ with $(\mu(0), y(0))) = (\mu_0, \xi_0)$. Since $y^*(\iota\xi(s)) = 1$, the choice of y^* gives that $\xi(s) = \xi_0 + \eta(s)$ where $\iota\eta(s) \in \text{range}(\mu_0\iota - L(0))$. The result of the proposition will have been proven once it is shown that each $\mu(s)$ is a simple eigenvalue of $L(s)$.

It follows from Proposition 2.7.7 that, for s sufficiently small, $\mu(s)\iota - L(s)$ is a Fredholm operator of index zero. Suppose that for $s \in (-\epsilon, \epsilon)$ there exists $x(s)$ such that

$$(\mu(s)\iota - L(s))x(s) = 0, \quad \|x(s)\|_X = 1.$$

Let $x(s) = \alpha(s)\xi_0 + z(s)$ where, as in Lemma 2.7.9, $z(s) \in X_1 = \iota^{-1}(\text{range}(\mu_0\iota - L(0)))$ and without loss of generality suppose $\alpha(s) \geq 0$. Then

$$(\mu_0\iota - L(0))z(s) = (\mu_0\iota - L(0))x(s)$$
$$= \left((\mu_0\iota - \mu(s))\iota - (L(0) - L(s))\right)x(s)$$
$$\to 0 \text{ in } Y \text{ as } s \to 0.$$

Therefore $z(s) \to 0$ in X, and so $\alpha(s) \to \|\xi_0\|_X^{-1}$, as $s \to 0$. Now let $\hat{x}(s) = x(s)/\alpha(s)$. Then $(\mu(s), \hat{x}(s)) \to (\mu_0, \xi_0)$ in $\mathbb{R} \times X$ and $G(\mu(s), \hat{x}(s), s) = 0$. The implicit function theorem 3.5.4 implies that, for s sufficiently small, $x(s)$ is a scalar multiple of $\xi(s)$. This shows that $\ker(\mu(s)\iota - L(s)) = \text{span}\{\xi(s)\}$.

Since $\xi(s) \to \xi_0$ in X and $\xi_0 \notin X_1 = \iota^{-1}(Y_1)$, which is closed in X, it follows that $\xi(s) \notin X_1$ for all s sufficiently small. Therefore $X = X_1 \oplus \text{span}\{\xi(s)\}$ for s sufficiently small.

Finally suppose that for $s \in (-\epsilon, \epsilon)$ there exists $p(s) \in X$ such that

$$(\mu(s)\iota - L(s))p(s) = \iota\xi(s).$$

By the preceding paragraph there is no loss in assuming that $p(s) \in X_1$. If $\|p(s_n)\|_X \to \infty$ for a sequence $s_n \to 0$, it is immediate that

$$\left(\mu_0 \iota - L(0)\right)\left(\frac{p(s_n)}{\|p(s_n)\|_X}\right) \to 0 \text{ in } Y$$

as $n \to \infty$. Therefore $p(s_n)/\|p(s_n)\|_X \to 0$ in X, by Lemma 2.7.9. But this is false. Hence $\|p(s)\|_X$ is bounded for s sufficiently small. Therefore

$$\left(\mu_0 \iota - L(0)\right)p(s) = \left((\mu_0 - \mu(s))\iota - (L(0) - L(s))\right)p(s) + \iota\xi(s) \to \iota\xi_0$$

in Y. Since $Y_1 = \mathrm{range}\,(\mu_0\iota - L(0))$ is closed in Y, $\iota\xi_0 \in Y_1$. This contradiction proves that $\mu(s)$ is a simple eigenvalue of $L(s)$. $\qquad\square$

3.7 NOTES ON SOURCES

Calculus in Banach spaces is covered in many textbooks, for example Cartan [16], Dieudonné [28] and Schwartz [52].

Chapter Four

Multilinear and Analytic Operators

The theory of higher order Fréchet derivatives leads to the notion of multilinear operators. The question of whether the Taylor polynomials of F (see Definition 3.3.2) converge to $F(x)$ for some $x \in X$ as $n \to \infty$ leads to the theory of analytic functions.

4.1 BOUNDED MULTILINEAR OPERATORS

Suppose that Y and $X_1, \cdots, X_p$, $p \in \mathbb{N}$, are Banach spaces over $\mathbb{F}$. A mapping $m : X_1 \times \cdots \times X_p \to Y$ is said to be a multilinear operator, in this case p-linear, if it is linear in each variable separately, that is, for all $k \in \{1, \cdots, p\}$ and $x_j \in X_j$, $j \neq k$,

$$x \mapsto m(x_1, \cdots, x_{k-1}, x, x_{k+1}, \cdots, x_p) \text{ is linear in } x \text{ over } \mathbb{F}.$$

It is said to be a bounded multilinear operator if, in addition,

$$\sup\{\|m(x_1, x_2, \cdots, x_p)\| : \|x_1\|, \cdots, \|x_p\| \leq 1\} = M < \infty. \qquad (4.1)$$

If m is multilinear and $x_j = 0$ for some j, then $m(x_1, \cdots, x_p) = 0$. Otherwise, if $x_j \neq 0$ for all j and m is bounded, we find that

$$\left\| m\left(\frac{x_1}{\|x_1\|}, \frac{x_2}{\|x_2\|}, \cdots, \frac{x_p}{\|x_p\|}\right) \right\| \leq M,$$

whence

$$\|m(x_1, \cdots, x_p)\| \leq M\|x_1\|\|x_2\| \cdots \|x_p\|.$$

The proofs of the next few propositions are so similar to those for $\mathcal{L}(X, Y)$ that we omit them.

PROPOSITION 4.1.1 *Suppose that $m : X_1 \times \cdots \times X_p \to Y$ is a multilinear operator. Then the following are equivalent statements.*

> *$m : X_1 \times \cdots \times X_p \to Y$ is continuous;*
>
> *m is continuous at $(0, \cdots, 0) \in X_1 \times \cdots \times X_p$;*
>
> *m is bounded.*

A bounded multilinear operator is called symmetric if

$$m(x_1, \cdots, x_p) = m(x_{\pi(1)}, \cdots, x_{\pi(p)})$$

for all $\pi \in S_p$, the symmetric group.

PROPOSITION 4.1.2   *The set $\mathcal{M}(X_1, \cdots, X_p; Y)$ of bounded multilinear operators endowed with the norm*

$$\|m\| = \sup\{\|m(x_1, x_2, \cdots, x_p)\| : \|x_1\|, \cdots, \|x_p\| \le 1\}$$

is a Banach space in which the symmetric operators form a closed subspace. When $X_k = X$, $1 \le k \le p$, we abbreviate $\mathcal{M}(X_1, \cdots, X_p; Y)$ as $\mathcal{M}^p(X, Y)$.

EXAMPLE 4.1.3   Fréchet derivatives give an important class of multilinear operators. If $F : U \subset X \to Y$ has a kth Fréchet derivative at $x_0 \in U$, it is clear (Proposition 3.2.3) that $(x_1, \cdots, x_k) \mapsto d^k F[x_0](x_1, \cdots, x_k)$ is a bounded, symmetric k-linear operator. That $d^k F : U \to \mathcal{M}^k(X, Y)$ is continuous is equivalent to saying that $F \in C^k(U, Y)$.   □

EXAMPLE 4.1.4   When $X = Y = \mathbb{F}$, it follows from the Riesz representation theorem 2.3.3 that every element m_p of $\mathcal{M}^p(X, Y)$ is given by

$$m_p(x_1, \cdots, x_p) = A_p \underbrace{x_1 \cdots x_p}_{\text{product}}$$

for some $A_p \in \mathbb{F}$, and therefore all m_p are symmetric in this case.   □

EXAMPLE 4.1.5   **(Determinants)** An $n \times n$ matrix A has rows $(a_{i1}, \cdots, a_{in}) \in \mathbb{F}^n$, $1 \le i \le n$. Its determinant $\det A$, defined by

$$\det A = \sum_{\pi \in S_n} \sigma(\pi) \prod_{i=1}^{n} a_{i\pi(i)}, \tag{4.2}$$

where $\sigma(\pi)$ denotes the signature of $\pi \in S_n$ (the symmetric group), is therefore an n-linear function on $(\mathbb{F}^n)^n$. Since interchanging two rows of A changes the sign of $\det A$, the determinant is not a symmetric operator on the rows of A. Instead we say that it is skew-symmetric. As a consequence, the determinant of a matrix with two equal rows is zero.

A function h on a Banach space X is called p-homogeneous if $f(\alpha x) = \alpha^p f(x)$ for all $\alpha \in \mathbb{F}$ and $x \in X$. Thus $A \mapsto \det A$ is n-homogeneous on the space of $n \times n$ matrices.   □

Let X^j denote the product of j copies of the same space X. When $m \in \mathcal{M}^p(X, Y)$ is symmetric, define a mapping on X^j by

$$(x_1 \cdots, x_j) \mapsto m(\underbrace{x_1, \cdots, x_1}_{k_1 \text{ times}}, \underbrace{x_2, \cdots, x_2}_{k_2 \text{ times}}, \cdots, \underbrace{x_j, \cdots, x_j}_{k_j \text{ times}})$$

$$= m\, x_1^{k_1} \cdots x_j^{k_j}, \tag{4.3}$$

where $j \in \mathbb{N}$ and $k_1 + \cdots + k_j = p$. (This defines the notation on the second line.)

PROPOSITION 4.1.6 *(a) Suppose $m \in \mathcal{M}(X_1, \cdots, X_p; Y)$. Then m is infinitely differentiable on the product space $X_1 \times \cdots \times X_p$ and its first derivative at a point $(x_1, \cdots, x_p) \in X_1 \times \cdots \times X_p$ is given by*

$$dm[(x_1, \cdots, x_p)](z_1, \cdots, z_p)$$
$$= \sum_{k=1}^{p} m(x_1, \cdots, x_{k-1}, z_k, x_{k+1}, \cdots, x_p),$$

for all $(z_1, \cdots, z_p) \in X_1 \times \cdots \times X_p$.
(b) If $m \in \mathcal{M}^p(X, Y)$ is symmetric then the mapping, h say, from X^j to Y defined by (4.3) is infinitely differentiable and for $(y_1, \cdots, y_j) \in X^j$,

$$dh[(x_1, \cdots, x_j)](y_1, \cdots, y_j)$$
$$= \sum_{l=1}^{j} k_l \, m x_1^{k_1} \cdots x_{l-1}^{k_{(l-1)}} x_l^{k_l - 1} x_{l+1}^{k_{(l+1)}} \cdots x_j^{k_j} \, y_l.$$

Proof. The proof, which is almost obvious and complicated only by notation, is left as an exercise. $\qquad\square$

REMARKS 4.1.7 The mapping

$$u_l \mapsto m x_1^{k_1} \cdots x_{l-1}^{k_{(l-1)}} x_l^{k_l - 1} x_{l+1}^{k_{(l+1)}} \cdots x_j^{k_j} \, u_l$$

belongs to $\mathcal{L}(X, Y)$ and so $m x_1^{k_1} \cdots x_{l-1}^{k_{(l-1)}} x_l^{k_l - 1} x_{l+1}^{k_{(l+1)}} \cdots x_j^{k_j}$ may be identified with an element of $\mathcal{L}(X, Y)$, which yields a shorthand notation for the partial derivative of h in part (b),

$$\partial_{x_l} h[(x_1, \cdots, x_j)] = k_l \, m x_1^{k_1} \cdots x_{l-1}^{k_{(l-1)}} x_l^{k_l - 1} x_{l+1}^{k_{(l+1)}} \cdots x_j^{k_j}.$$

$\qquad\square$

PROPOSITION 4.1.8 *Suppose that $m \in \mathcal{M}^p(X, Y)$ and define $F : X \to Y$ by*

$$F(x) = m(x, \cdots, x), \quad x \in X.$$

Then $d^k F[0] = 0$ except when $k = p$ and

$$d^p F[0](x_1, x_2, \cdots, x_p) = \sum_{\pi \in S_p} m(x_{\pi(1)}, x_{\pi(2)}, \cdots, x_{\pi(p)}),$$

where S_p is the symmetric group.

Proof. This is an elementary exercise using induction on the order of differentiation. $\qquad\square$

4.2 FAÀ DE BRUNO FORMULA

Although the formula for the kth derivative of the composition of two C^k-functions looks intimidating, its derivation is straightforward except that the notation is elaborate. Let X, Y and Z be Banach spaces with $U \subset X$ and $V \subset Y$ open sets. Suppose that $F \in C^k(U, Y)$ maps U into V, $G \in C^k(V, Z)$ and $H \in C^k(U, Z)$ is defined by

$$H(x) = G(F(x)), \quad x \in U.$$

The formula for $d^k H$ is most easily expressed in terms of some auxiliary notation. Let $(n+)$ denote the set of n-tuples of positive integers. For $\beta = (\beta_1, \cdots, \beta_n) \in (n+)$, let $\beta! = \beta_1! \beta_2! \cdots \beta_n!$ and $|\beta| = \beta_1 + \cdots + \beta_n$.

For $x_0 \in U$, $\beta \in (n+)$, $(x_1, \cdots, x_k) \in X^k$, let $\mathcal{D}_n^k(F, G)[x_0] : X^k \to Y$ be defined by

$$\mathcal{D}_n^k(F, G)[x_0](x_1, \cdots, x_k) =$$
$$\sum_{\substack{\sigma \in S_k \\ \beta \in (n+) \\ |\beta| = k}} \frac{1}{\beta!} d^n G[F(x_0)] \left(d^{\beta_1} F[x_0](x_{\sigma(1)}, \cdots, x_{\sigma(\beta_1)}), \right.$$
$$d^{\beta_2} F[x_0] \left(x_{\sigma(\beta_1+1)}, \cdots, x_{\sigma(\beta_1+\beta_2)} \right), \cdots$$
$$\left. \cdots, d^{\beta_n} F[x_0] \left(x_{\sigma(\beta_1+\beta_2+\cdots+\beta_{n-1}+1)}, \cdots, x_{\sigma(k)} \right) \right),$$

where S_k is the symmetric group.

THEOREM 4.2.1 **(Faà de Bruno Formula)** *For $x_0 \in U$ and $k \in \mathbb{N}$,*

$$d^k H[x_0](x_1, \cdots, x_k) = \sum_{n=1}^{k} \frac{1}{n!} \mathcal{D}_n^k(F, G)[x_0](x_1, \cdots, x_k).$$

Proof. Since both F and G have k^{th} Fréchet derivatives, so has H. From the Chain rule 3.1.3, it is obvious that the kth derivative of H at x_0 depends only on the first k derivatives of F at x_0 and of G at $F(x_0)$. Therefore it suffices to replace F by its kth order Taylor polynomial at x_0 and G by its kth order Taylor polynomial at $F(x_0)$, and without loss of generality we may assume that $x_0 = 0 \in X$, $F(x_0) = F(0) = 0 \in Y$ and $G(0) = 0 \in Z$. Thus henceforth

$$F(x) = \sum_{m=1}^{k} \frac{1}{m!} f_m x^m \quad \text{and} \quad G(y) = \sum_{n=1}^{k} \frac{1}{n!} g_n y^n$$

where $f_m = d^m F[0]$ and $g_n = d^n G[0]$ are bounded symmetric m-linear operators.

Therefore

$$H(x) = \sum_{n=1}^{k} \frac{1}{n!} g_n \left(\sum_{m=1}^{k} \frac{1}{m!} f_m x^m \right)^n$$

$$= \sum_{n=1}^{k} \frac{1}{n!} g_n \left(\sum_{\beta_1=1}^{k} \frac{1}{\beta_1!} f_{\beta_1} x^{\beta_1}, \sum_{\beta_2=1}^{k} \frac{1}{\beta_2!} f_{\beta_2} x^{\beta_2}, \cdots, \sum_{\beta_n=1}^{k} \frac{1}{\beta_n!} f_{\beta_n} x^{\beta_n} \right).$$

By Proposition 4.1.8, in this expression for $H(x)$ the k-linear term

$$\sum_{n=1}^{k} \frac{1}{n!} \left(\sum_{\substack{\beta \in (n+) \\ |\beta|=k}} g_n \left(\frac{1}{\beta_1!} f_{\beta_1} x^{\beta_1}, \frac{1}{\beta_2!} f_{\beta_2} x^{\beta_2}, \cdots, \frac{1}{\beta_n!} f_{\beta_n} x^{\beta_n} \right) \right)$$

$$= \sum_{n=1}^{k} \frac{1}{n!} \left(\sum_{\substack{\beta \in (n+) \\ |\beta|=k}} \frac{1}{\beta!} d^n G[F(x_0)] \left(d^{\beta_1} F[x_0] x^{\beta_1}, \cdots, d^{\beta_n} F[x_0] x^{\beta_n} \right) \right),$$

is a function of $x \in X$ whose k^{th} derivative everywhere coincides with the kth derivative of H at 0. A further application of Proposition 4.1.8 now leads to the Faà de Bruno formula. This completes the proof. $\square$

4.3 ANALYTIC OPERATORS

Now we embark on a study of power series and $\mathbb{F}$- analytic functions in Banach spaces. Let X and Y be Banach spaces over $\mathbb{F}$. Let U be an open subset of X.

DEFINITION 4.3.1   *A mapping $F : U \to Y$ is $\mathbb{F}$- analytic at $x_0 \in U$ if, for all $x \in U$ with $\|x - x_0\|$ sufficiently small,*

$$F(x) = \sum_{k=0}^{\infty} m_k (x - x_0)^k \tag{4.4}$$

where $F(x_0) = m_0(x - x_0)^0 = m_0 \in Y$, $m_k \in \mathcal{M}^k(X, Y)$ is symmetric and there exists $r > 0$ such that

$$\sup_{k \geq 0} r^k \|m_k\| = M < \infty. \tag{4.5}$$

The series on the right in (4.4) is a power series in $x - x_0$. The function F is said to be $\mathbb{F}$-analytic on U if it is $\mathbb{F}$-analytic at every point of U. (When it does not matter we will omit $\mathbb{F}$ and speak of analytic functions.) The expressions analytic mapping, analytic operator and analytic function will be used interchangeably in what follows.

Because of (4.5), for all $x \in X$ with $\|x - x_0\| < r$,

$$\sum_{k=0}^{\infty} \|m_k(x - x_0)^k\| \leq M \sum_{k=0}^{\infty} \frac{\|x - x_0\|^k}{r^k} = \frac{Mr}{r - \|x - x_0\|} < \infty$$

and therefore the sequence $\left\{ \sum_{k=0}^{n} m_k(x - x_0)^k \right\}$ of partial sums is summable in norm. For an open set $U \subset X$ let $C^{\infty}(U, Y)$ denote the set of all functions $F : U \to Y$ which have derivatives of all orders at every point of U,

$$C^{\infty}(U, Y) = \cap_{k=1}^{\infty} C^k(U, Y).$$

If $F \in C^{\infty}(U, Y)$ and $x_0, x \in U$ then we have seen in the proof of Taylor's theorem 3.3.1 that

$$F(x) - \sum_{k=0}^{n} \frac{1}{k!} d^k F[x_0](x - x_0)^k = R_n(x, x_0) \tag{4.6}$$

where

$$\|R_n(x, x_0)\| \leq \frac{\|x - x_0\|^{n+1}}{(n+1)!} \sup_{0 \leq t \leq 1} \|d^{n+1} F[(1 - t)x_0 + tx]\|.$$

Here $R_n(\cdot, x_0)$ is called the remainder after n terms in the Taylor expansion of F at x_0. Since $dF^k[x_0] \in \mathcal{M}^k(X, Y)$ is symmetric for all $k \in \mathbb{N}$ when $F \in C^{\infty}(U, Y)$, its analyticity at $x_0 \in U$ is equivalent (Proposition 4.3.4) to the convergence to zero of the remainder at every point x in a neighbourhood of x_0.

THEOREM 4.3.2 *Suppose that $U \subset X$ is open and $F \in C^{\infty}(U, Y)$. Suppose also that for each $x_0 \in U$ there exist constants $r, C, R > 0$, all depending on x_0, such that*

$$\left\| d^k F[x] \right\| \leq \frac{C\, k!}{R^k} \text{ for all } x \in U \text{ with } \|x - x_0\| < r. \tag{4.7}$$

Then F is analytic on U.

REMARK 4.3.3 Since Stirling's formula says that

$$\lim_{n \to \infty} \frac{\sqrt{2\pi n}\, n^n}{e^n\, n!} = 1, \tag{4.8}$$

(4.7) is equivalent to

$$\left\| d^k F[x] \right\| \leq \frac{C\, k^k}{R^k} \text{ for all } x \in U \text{ with } \|x - x_0\| < r,$$

for different constants C and R. (The inequality $e^n \geq n^n/n!$ is sufficient for this observation.) $\qquad\qquad\Box$

Proof. Let x_0, $x \in U$. Then in (4.6),

$$\|R_n(x, x_0)\| \leq \frac{\|x - x_0\|^{n+1}}{(n+1)!} \sup_{0 \leq t \leq 1} \|d^{n+1} F[(1-t)x_0 + tx]\|$$

$$\leq \frac{C}{R^{n+1}} \|x - x_0\|^{n+1} \to 0 \text{ as } n \to \infty$$

if $\|x - x_0\| < \min\{r, R\}$. This proves the result. $\qquad\square$

We aim to prove a converse of Theorem 4.3.2, but first some observations. Differentiating the identity

$$\frac{1}{1-x} = \sum_{k=0}^{\infty} x^k, \quad |x| < 1,$$

p times yields the new identity

$$\left(\frac{1}{1-x}\right)^{p+1} = \sum_{k=0}^{\infty} \binom{k+p}{p} x^k, \quad |x| < 1, \ p \in \mathbb{N}. \qquad (4.9)$$

Suppose that $r > 0$ and $\{m_k\}$ is a sequence of symmetric k-linear operators on X with $r^k \|m_k\| \leq M$ for all $k \in \mathbb{N}$. Let $x, z_1, \cdots, z_p \in X$ with $\|x\| < (1-\epsilon)r$, $\epsilon \in (0, 1)$, and note that, by (4.9),

$$\sum_{k=0}^{\infty} \binom{k+p}{p} \|m_{k+p} x^k z_1 \cdots z_p\|$$

$$\leq \sum_{k=0}^{\infty} \binom{k+p}{p} \|m_{k+p}\| \|x\|^k \|z_1\| \cdots \|z_p\|$$

$$\leq M \frac{\|z_1\| \cdots \|z_p\|}{r^p} \sum_{k=0}^{\infty} \binom{k+p}{p} \left(\frac{\|x\|}{r}\right)^k$$

$$= M \frac{\|z_1\| \cdots \|z_p\|}{r^p} \left(\frac{r}{r - \|x\|}\right)^{p+1}$$

$$\leq \epsilon^{-1} M \left(\frac{\|z_1\| \cdots \|z_p\|}{\epsilon^p r^p}\right).$$

Hence, when $\|x\| < (1-\epsilon)r$, a symmetric operator $M_p^x \in \mathcal{M}^p(X, Y)$ may be defined at $(z_1, \cdots, z_p) \in X^p$ by

$$M_p^x(z_1, \cdots, z_p) = \sum_{k=0}^{\infty} \binom{k+p}{p} m_{k+p} x^k z_1 \cdots z_p \in Y$$

and

$$\|M_p^x\| \leq \frac{M}{r^p \epsilon^{p+1}}. \qquad (4.10)$$

Also, if $\|x\| < (1 - \epsilon)r$ and $\|z\| < \epsilon r$, then

$$\sum_{k,p=0}^{\infty} \binom{k+p}{p} \|m_{k+p} x^k z^p\| \leq \frac{M}{\epsilon} \sum_{p=0}^{\infty} \left(\frac{\|z\|}{\epsilon r}\right)^p < \infty, \tag{4.11}$$

and so the series

$$\sum_{k,p=0}^{\infty} \binom{k+p}{p} m_{k+p} x^k z^p$$

is summable in norm, and therefore convergent to a sum which is independent of the order in which summation is done. This leads to the following result in which (4.13) is a converse of Theorem 4.3.2.

PROPOSITION 4.3.4 *Let (4.5) hold and let F be defined by (4.4). Then F is analytic at every point x of the set $U_0 = \{x \in U : \|x - x_0\| < r\}$,*

$$F \in C^{\infty}(U_0, Y) \text{ and } m_k = \frac{d^k F[x_0]}{k!}, \text{ for all } k \geq 0.$$

Also the pth derivative, $d^p F$, is analytic at x for all $p \in \mathbb{N}$ and $\|x - x_0\| < r$,

$$d^p F[x](x_1, \cdots, x_p) = \sum_{k=0}^{\infty} \frac{(k+p)!}{k!} m_{k+p}(x - x_0)^k x_1 x_2 \cdots x_p, \tag{4.12}$$

and there exists $C > 1$, $R \in (0, 1)$ such that if $\|x - x_0\| \leq \frac{1}{2}r$,

$$\|d^p F[x]\| \leq C \frac{p!}{R^p} \text{ for all } p \in \mathbb{N}. \tag{4.13}$$

If $K \subset U$ is a compact set, then there exists C and R such that (4.13) holds for all $x \in K$.

Proof. We will show first that if $\|\hat{x} - x_0\| < r$ then F is analytic at $\hat{x}$. This follows from (4.11) since

$$F(x) = \sum_{l=0}^{\infty} m_l (x - x_0)^l = \sum_{l=0}^{\infty} m_l \left((x - \hat{x}) + (\hat{x} - x_0)\right)^l$$

$$= \sum_{l=0}^{\infty} \sum_{k=0}^{l} \binom{l}{k} m_l (x - \hat{x})^{l-k} (\hat{x} - x_0)^k$$

$$= \sum_{k=0}^{\infty} \sum_{l=k}^{\infty} \binom{l}{k} m_l (x - \hat{x})^{l-k} (\hat{x} - x_0)^k$$

$$= \sum_{k=0}^{\infty} \sum_{p=0}^{\infty} \binom{k+p}{k} m_{p+k} (x - \hat{x})^p (\hat{x} - x_0)^k$$

$$= \sum_{p=0}^{\infty} M_p^{\hat{x}-x_0} (x - \hat{x})^p \in Y. \tag{4.14}$$

Since (4.10) holds, this shows that F is analytic at $\hat{x}$ provided $\|\hat{x} - x_0\| < r$.

Next, observe that when $\|x - x_0\| < r$,

$$\|F(x) - F(x_0) - m_1(x - x_0)\| = \left\| \sum_{l=2}^{\infty} m_l(x - x_0)^l \right\| = o(\|x - x_0\|)$$

as $\|x - x_0\| \to 0$, by (4.5). Therefore F is Fréchet differentiable at x_0 and

$$dF[x_0]x = m_1 x \quad \text{for all } x \in X.$$

By the same token it follows from (4.14) that when $\|\hat{x} - x_0\| < r$, F is differentiable at $\hat{x}$ and

$$dF[\hat{x}]x = M_1^{(\hat{x} - x_0)} x = \sum_{k=0}^{\infty}(k+1)m_{k+1}(\hat{x} - x_0)^k x.$$

Thus dF is analytic at x_0 since, for x with $\|x - x_0\|$ sufficiently small,

$$dF[x] = \sum_{k=0}^{\infty}(k+1)m_{k+1}(x - x_0)^k$$

where the right-hand side is regarded as an element of $\mathcal{L}(X,Y)$ expressed as a power series. (To confirm that (4.5) holds here note that

$$(k+1)\|m_{k+1}\| \leq M(k+1)r^{k+1} \leq \{M/r\}(2/r)^k.$$

Now replace r with $r/2$ and M with M/r.)

It now follows by induction that all the derivatives of F exist at every point of x with $\|x - x_0\| < r$ and are given by a power series in $x - x_0$. The formula (4.12) also follows by induction. Therefore for $p \in \mathbb{N}$ and $x \in X$ with $\|x - x_0\| \leq \frac{1}{2}r$,

$$\|d^p F[x]\| \leq \sum_{k=0}^{\infty} \frac{(k+p)!}{k!}\|m_{k+p}\| \, \|x - x_0\|^k$$

$$\leq \sum_{k=0}^{\infty} \frac{(k+p)!}{k!} \frac{M}{r^{k+p}}\left(\frac{r}{2}\right)^k = M\frac{p!}{r^p} \sum_{k=0}^{\infty} \frac{(k+p)!}{k!\,p!}\left(\frac{1}{2}\right)^k$$

$$= 2M\frac{2^p\,p!}{r^p} \quad \text{by (4.9).}$$

Finally, if $K \subset U$ is compact then (4.13) holds uniformly for $x \in K$, since every open cover of K has a finite sub-cover. This completes the proof. $\quad\square$

Here are important, non-trivial examples of analytic operators.

EXAMPLE 4.3.5 (**Operator Inverses**) Suppose $T \in \mathcal{L}(X,Y)$ is a bijection. Then, by Lemma 2.5.1, there exists $\epsilon > 0$ such that if $L \in \mathcal{L}(X,Y)$ and $\|L - T\| < \epsilon$, then $L^{-1} \in \mathcal{L}(Y,X)$ and

$$L^{-1} = \left(I - T^{-1}(T - L)\right)^{-1}T^{-1} = \sum_{k=0}^{\infty}\left(T^{-1}(T - L)\right)^k T^{-1}.$$

Let a symmetric $m_k \in \mathcal{M}^k\big(\mathcal{L}(X,Y), \mathcal{L}(Y,X)\big)$ be defined by

$$m_k(L_1, \cdots, L_k)$$
$$= \frac{1}{k!} \sum_{\pi \in S_k} T^{-1} \circ L_{\pi(1)} \circ T^{-1} \circ L_{\pi(2)} \circ \cdots \circ T^{-1} \circ L_{\pi(k)} \circ T^{-1}.$$

Then

$$L^{-1} = \sum_{k=0}^{\infty} m_k(T - L)^k,$$

which shows that $L \mapsto L^{-1}$ from $\mathcal{L}(X,Y)$ to $\mathcal{L}(Y,X)$ is $\mathbb{F}$-analytic on the open set where it is defined. $\qquad\square$

EXAMPLE 4.3.6 **(Analytic Nemytskii Operators)** Let $f : \mathbb{F}^N \to \mathbb{F}^M$ be $\mathbb{F}$-analytic. Then the Nemytskii operator F defined in Example 3.1.9 is $\mathbb{F}$-analytic on $X := C\big([0,1], \mathbb{F}^N\big)$. (Note that this is not the same as saying that the composition of two analytic functions is analytic.) To see that F is $\mathbb{F}$-analytic at $u_0 \in X$ we proceed as follows. Let $M = \sup\{|u_0(t)| : t \in [0,1]\}$. Let $C > 1$ and $R \in (0,1)$, given by the last part of Proposition 4.3.4, be such that

$$\|f^k[\xi]\| \leq \frac{Ck!}{R^k} \quad \text{for } \xi \in \mathbb{F}^N \text{ with } |\xi| \leq M + 1. \tag{4.15}$$

Now for $u \in X$ with $\|u\| \leq M + 1$, $t \in [0,1]$ and $n \in \mathbb{N}$ let

$$f_n(u(t)) = \sum_{k=0}^{n} \frac{f^{(k)}(u_0(t))}{k!}(u(t) - u_0(t))^k.$$

By Taylor's theorem,

$$f(u(t)) - f_n(u(t)) = R_n(u(t), u_0(t))$$

where

$$\|R_n(u(t), u_0(t))\|$$
$$\leq \frac{\|u(t) - u_0(t)\|^{k+1}}{(k+1)!} \sup_{0 \leq s \leq 1} \|f^{(k+1)}((1-s)u_0(t) + su(t))\|.$$

It follows from (4.15) that $f_n(u) \to f(u)$ in the Banach space X as $n \to \infty$ provided that $\|u - u_0\| < R$. It remains to define symmetric $m_k \in \mathcal{M}^k(X,X)$ by

$$m_k(v_1, \cdots, v_k)(t) = \frac{d^k f[u_0(t)]}{k!}(v_1(t), v_2(t), \cdots, v_k(t))$$

so that

$$F(u) = \sum_{k=0}^{\infty} m_k(u - u_0)^k,$$

where the series on the right converges in $C([0,1], \mathbb{F}^M)$. Therefore the Nemytskii operator is $\mathbb{F}$-analytic on X. A similar argument leads to the same conclusion for the same Nemytskii operator regarded as a mapping on $C^n([0,1], \mathbb{F}^N)$, and on other Banach spaces of functions as well. $\qquad \square$

REMARK 4.3.7 **(Analytic Operators on Lebesgue Spaces)** Suppose P is a polynomial of degree p and $P \circ u \in L_q[0,1]$ for all $u \in L_r[0,1]$. It is easy to see that $pq \leq r$.

Consider the mapping $f(u) = u^p$ where $p \in \mathbb{N}$ and $pq \leq r$. It follows that the kth-derivative, $k \leq p$, of f at $u_0 \in L_r[0,1]$ is given by

$$d^k f[u_0](u_1, \cdots, u_k) = \frac{p!}{(p-k)!} u_0^{p-k} u_1 \cdots u_k,$$

which, by Hölder's inequality, is a bounded k-linear operator on $L_r[0,1]$. All higher derivatives of f are zero. The Nemytskii operator from $L_r[0,1]$ to $L_q[0,1]$ so defined is then $\mathbb{F}$-analytic. For example $u \mapsto u^2$ is analytic from $L_3[0,1]$ to $L_1[0,1]$. (However there are no non-trivial $\mathbb{F}$-analytic operators from $L_p[0,1]$ to itself, $1 \leq p < \infty$, as Example 3.1.10 shows.) $\qquad \square$

It is well known that if $U \subset \mathbb{C}$ is open and connected, and $f : U \to \mathbb{C}$ is a non-constant $\mathbb{C}$-analytic function, then the zero set of f has no limit points in U. This is clearly false for $\mathbb{C}$-analytic functions from $\mathbb{C}^2$ into $\mathbb{C}^2$, as the following example shows.

EXAMPLE 4.3.8 Let $F : \mathbb{C}^2 \to \mathbb{C}^2$ be defined by

$$F(x, y) = (xy, xy)$$

for all $(x, y) \in \mathbb{C}^2$. Then F is $\mathbb{C}$-analytic because its Taylor series has one term: $F(x, y) = m_2(x, y)^2$ where, for $(x_1, y_1), (x_2, y_2) \in \mathbb{C}^2$,

$$m_2((x_1, y_1), (x_2, y_2)) = \tfrac{1}{2}(x_1 y_2 + x_2 y_1, x_1 y_2 + x_2 y_1).$$

However $F(x, y) = (0, 0)$ when $xy = 0$. Thus every point of the zero set of F is a limit point of the zero set. $\qquad \square$

However we have the following result.

THEOREM 4.3.9 *Suppose that X and Y are Banach spaces, that $U \subset X$ is an open connected set and that $F : U \to Y$ is $\mathbb{F}$-analytic. Suppose also that there is a non-empty open set $W \subset U$ on which F is identically 0. Then F is identically zero on U.*

Proof. Let $V \subset U$ be the set of points $x \in U$ at which all the derivatives of F are zero. The set V is non-empty since, by hypothesis, $W \subset V$. Since $F \in C^\infty(U, Y)$, V is an intersection of sets which are closed in U and so V is closed in U. Since F is an $\mathbb{F}$-analytic function, Proposition 4.3.4 implies that V is open, and therefore open in U. But U is connected, so $U = V$. This completes the proof. $\qquad \square$

4.4 ANALYTIC FUNCTIONS OF TWO VARIABLES

Let X, Y and Z be Banach spaces and U an open subset of $X \times Y$. Let $(x_0, y_0) \in U$ and let $F : U \to Z$ be analytic at (x_0, y_0). In other words, for (x, y) sufficiently close to (x_0, y_0),

$$F(x, y) = \sum_{k=0}^{\infty} m_k (x - x_0, y - y_0)^k \qquad (4.16)$$

where $m_k \in \mathcal{M}^k(X \times Y, Z)$ is symmetric with $\sup_{k \geq 0} r^k \|m_k\| < \infty$ for some $r > 0$. Now let

$$\mathcal{M}^{p,q}(X, Y; Z) = \mathcal{M}(\underbrace{X, \cdots, X}_{p \text{ times}}, \underbrace{Y, \cdots, Y}_{q \text{ times}}; Z),$$

and define $m_{p,q} \in \mathcal{M}^{p,q}(X, Y; Z)$ by

$$m_{p,q}(x_1, \cdots, x_p, y_1, \cdots, y_q)$$
$$= m_{p+q}\big((x_1, 0), \cdots, (x_p, 0), (0, y_1), \cdots, (0, y_q)\big).$$

It follows that $\|m_{p,q}\| \leq \|m_{p+q}\|$ and, since $(x, y) = (x, 0) + (0, y)$, an argument similar to that for (4.14) now yields that

$$F(x, y) = \sum_{p, q \geq 0} \frac{(p+q)!}{p! \, q!} m_{p,q}((x - x_0)^p, (y - y_0)^q), \qquad (4.17)$$

$$\sup_{p, q \geq 0} \big\{ \|m_{p, q}\| r^{p+q} \big\} < \infty, \qquad (4.18)$$

where $r > 0$. Note that $m_{p,q}$ is symmetric separately in the first p, and in the last q, variables. The series

$$\sum_{p, q \geq 0} \frac{(p+q)!}{p! \, q!} m_{p,q}((x - x_0)^p, (y - y_0)^q)$$

is summable in norm for $\|x - x_0\| + \|y - y_0\| < r$. It follows that

$$m_{p,q} = \frac{\partial_x^p \partial_y^q F[(x_0, y_0)]}{(p+q)!}$$

where $\partial_x^p \partial_y^q F[(x_0, y_0)] \in \mathcal{M}^{p,q}(X, Y; Z)$, and hence

$$F(x, y) = \sum_{p, q \geq 0} \frac{\partial_x^p \partial_y^q F[(x_0, y_0)]}{p! \, q!} ((x - x_0)^p, (y - y_0)^q), \qquad (4.19)$$

$$\sup_{p, q \geq 0} \Big\{ \frac{\|\partial_x^p \partial_y^q F[(x_0, y_0)]\| r^{p+q}}{(p+q)!} \Big\} < \infty, \qquad (4.20)$$

for some $r > 0$. Conversely, (4.17) and (4.18) defines an $\mathbb{F}$-analytic mapping, in the sense of Definition 4.3.1, which satisfies (4.19) and (4.20).

4.5 ANALYTIC INVERSE AND IMPLICIT FUNCTION THEOREMS

We begin by proving the analytic version of the inverse function theorem. Note that the proof here does not assume that the composition of two analytic functions is analytic; instead this emerges as a corollary of the theorem.

Let X be a Banach space and let $B_r(X)$, $r \in (0,1)$, denote the ball of radius r centred at $0 \in X$. For $r \in (0,1)$, let $\mathcal{B}_r = B_{r^2}(X) \times B_r(X) \subset X \times X$ and let $(E_r, \|\cdot\|_r)$ denote the Banach space of functions u which are analytic from $\mathcal{B}_r$ into X with

$$u(x,y) = \sum_{m,n \geq 0} u_{m,n} x^m y^n, \quad (x,y) \in X \times X,$$

where

$$\|u\|_r = \sum_{m,n \geq 0} \|u_{m,n}\| r^{2m+n} < \infty.$$

Note that the norm $\|\cdot\|_r$ uses different weights for the x and y dependence of functions in E_r. However, since (4.16) implies (4.17) and (4.18), *any* function F which maps a neighbourhood of the origin in $X \times X$ analytically into X belongs to E_r for all $r > 0$ sufficiently small. That E_r is a Banach space follows by a proof, almost identical to that of Proposition 5.1.3 (below), in which the completeness of $\mathbb{F}$ is replaced with the completeness (Proposition 4.1.2) of the space of k-linear operators on X for all $k \in \mathbb{N}_0$. Let F_r denote the closed subspace of E_r of functions $u \in E_r$ with

$$u(x,y) = \sum_{m \geq 0, n \geq 1} u_{m,n} x^m y^n.$$

Define $L \in \mathcal{L}(F_r, F_r)$ by

$$Lu(x,y) = \sum_{m \geq 0, n \geq 1} \frac{u_{m,n}}{n} x^m y^n, \quad (x,y) \in \mathcal{B}_r, \quad u \in F_r.$$

Clearly $\|L\| = 1$. Now for $w \in E_r$ arbitrary but fixed define $L_w u$ for $u \in F_r$ by

$$L_w u(x,y) = \partial_y u[(x,y)] w(x,y) - \partial_y u[(x,0)] w(x,0), \quad (x,y) \in \mathcal{B}_r.$$

LEMMA 4.5.1 *(a) The operator $L_w \circ L \in \mathcal{L}(F_r, F_r)$ and $\|L_w \circ L\| \leq \|w\|_r / r$. (b) Let $w_0 \in E_r$ be defined by $w_0(x,y) = y$, $(x,y) \in \mathcal{B}_r$. Then $L_{w_0} \circ L$ is the identity on F_r.*

Proof. Let $w(x, y) = \sum_{p,\,q\geq 0} w_{p,q} x^p y^q$. Then for $u \in F_r$,

$$L_w \circ Lu\,(x, y)$$

$$= \sum_{m,n\geq 0} u_{m,n+1} x^m y^n \left(\sum_{p,\,q\geq 0} w_{p,q} x^p y^q \right) - \sum_{m\geq 0} u_{m,1} x^m \left(\sum_{p\geq 0} w_{p,0} x^p \right)$$

$$= \sum_{\substack{M\geq 0,\, N\geq 1}} \sum_{\substack{m+p=M \\ n+q=N}} u_{m,n+1} x^m y^n \left(w_{p,q} x^p y^q \right),$$

from which it follows that

$$\| L_w \circ Lu \|_r \leq \sum_{M\geq 0,\, N\geq 1} r^{2M+N} \left(\sum_{\substack{m+p=M \\ n+q=N}} \| u_{m,n+1} \|\, \| w_{p,q} \| \right)$$

$$= \frac{1}{r} \sum_{M\geq 0,\, N\geq 1} \left(\sum_{\substack{m+p=M \\ n+q=N}} \left(r^{2m+n+1} \| u_{m,n+1} \| \right) \left(r^{2p+q} \| w_{p,q} \| \right) \right)$$

$$\leq \frac{1}{r} \left(\sum_{\substack{m\geq 0 \\ n\geq 0}} r^{2m+n+1} \| u_{m,n+1} \| \right) \left(\sum_{\substack{p\geq 0 \\ q\geq 0}} r^{2p+q} \| w_{p,q} \| \right)$$

$$= \frac{\| u \|_r \| w \|_r}{r} < \infty.$$

It now follows that $L_w \circ L \in \mathcal{L}(F_r, E_r)$ and $\| L_w \circ L \| \leq \| w \|_r / r$. Since $L_w \circ Lu\,(x, 0) = 0 \in X$, $L_w \circ L \in \mathcal{L}(F_r, F_r)$ and part (a) is proven.

(b) It is immediate from the definitions that $L_{w_0} \circ L$ is the identity operator on F_r. $\qquad\square$

PROPOSITION 4.5.2 *Suppose that F maps a neighbourhood of the origin in X analytically into X with $F(0) = 0$ and $dF[0] = I$. Then there exist open neighbourhoods V, U of the origin in X and an $\mathbb{F}$-analytic function $G : V \to X$ such that the following statements are equivalent:*

$$F(y) = x, \; y \in U \; \text{and} \; G(x) = y, \; x \in V.$$

Proof. Suppose that F is analytic from a neighbourhood of $0 \in X$ into X with $F(0) = 0$ and $dF[0] = I$. For $r > 0$ sufficiently small let v, $w \in E_r$ be defined for $(x, y) \in B_r$ by

$$v(x, y) = F(y) - x \; \text{and} \; w(x, y) = v(x, y) - w_0(x, y) = F(y) - x - y.$$

Then

$$w(x, y) = -x + \sum_{n\geq 2} \frac{d^n F[0]}{n!} y^n,$$

and

$$\|w\|_r \le r^2 + \sum_{n\ge 2} \frac{\|d^n F[0]\|}{n!} r^n \le r^2 C(F),$$

where $C(F)$ is a constant determined by F. From the definitions

$$L_v \circ L - I = (L_v - L_{w_0}) \circ L = L_w \circ L,$$

and hence, by the preceding lemma, $\|L_v \circ L - I\| \le r C(F)$ for $r > 0$ sufficiently small. Therefore, for $r > 0$ sufficiently small, $L_v \circ L$ is a homeomorphism on F_r by Lemma 2.5.1. So there exists $u_0 \in F_r$ such that $L_v \circ L u_0 = w_0$ and, for all $(x, y) \in B_r$,

$$L_v \circ L u_0\,(x, y) = \partial_y(L u_0)[(x, y)]v(x, y) - \partial_y(L u_0)[(x, 0)]v(x, 0) = y.$$

In particular, when $y \in X$ and $t \in (0, 1]$ is sufficiently small,

$$ty = L_v \circ L u_0\,(0, ty) = \partial_y(L u_0)[(0, ty)]\big(F(ty) - F(0)\big),$$

which (dividing by t and letting $t \to 0$) gives that $\partial_y(L u_0)[(0, 0)] = I$, the identity on X. Hence there exists ϵ with $0 < \epsilon < r$ such that if $(x, y) \in B_\epsilon$, then $\partial_y(L u_0)[(x, y)]$ is a bijection on X. Moreover, for $(x, y) \in B_\epsilon$,

$$\partial_y(L u_0)[(x, y)]\big(F(y) - x\big) = y - G(x), \tag{4.21}$$

where G is defined on $B_{\epsilon^2}(X)$ by

$$G(x) = -\partial_y(L u_0)[(x, 0)]v(x, 0) = \partial_y(L u_0)[(x, 0)]x.$$

It is clear that G is $\mathbb{F}$-analytic. Let $V = B_{\epsilon^2}(X) \cap G^{-1}(B_\epsilon(X))$ and $U = B_\epsilon(X) \cap F^{-1}(B_{\epsilon^2}(X))$, open neighbourhoods of $0 \in X$. Since $\partial_y(L u_0)[(x, y)]$ in (4.21) is a bijection, this completes the proof. $\qquad\square$

This result has four important corollaries.

THEOREM 4.5.3 (Analytic Inverse Function Theorem) *Suppose that X, Y are Banach spaces, that $x_0 \in U \subset X$, where U is open. Suppose also that $F : U \to Y$ is analytic and that $dF[x_0] \in \mathcal{L}(X, Y)$ is a homeomorphism. Then there exist opens sets U_0 and V_0 with $x_0 \in U_0 \subset U$, $F(x_0) \in V_0 \subset Y$ and an analytic map $G : V_0 \to X$ such that*

$$for\ x \in U_0, \quad F(x) \in V_0 \ and \ G(F(x)) = x,$$

and

$$for\ y \in V_0, \quad G(y) \in U_0 \ and \ F(G(y)) = y.$$

Proof. Let $\tilde{U} = U - x_0$, replace $F : U \to Y$ with $\tilde{F} : \tilde{U} \to X$ defined by

$$\tilde{F}(y) = (dF[x_0])^{-1}(F(y + x_0) - F(x_0)),$$

and apply the preceding proposition. □

THEOREM 4.5.4 (Analytic Implicit Function Theorem) *Let X, Y and Z be Banach spaces and suppose that $U \subset X \times Y$ is open. Let $(x_0, y_0) \in U$ and suppose that $F : U \to Z$ is analytic where the partial derivative $\partial_x F[(x_0, y_0)] \in \mathcal{L}(X, Z)$ is a homeomorphism.*

Then there exists an open neighbourhood $V \subset Y$ of y_0, an open set $W \subset U$ and an $\mathbb{F}$-analytic mapping $\phi : V \to X$ such that

$$(x_0, y_0) \in W \text{ and } F^{-1}(z_0) \cap W = \{(\phi(y), y) : y \in V\}.$$

Proof. In the light of Theorem 4.5.3 the proof is the same as that of Theorem 3.5.4. □

DEFINITION 4.5.5 *A set $M \subset \mathbb{F}^n$ is called an m-dimensional $\mathbb{F}$-analytic manifold if, for all points $a \in M$, there is an open neighbourhood U_a of $0 \in \mathbb{F}^m$ and an analytic function $f : U_a \to M$ such that $f(0) = a$, $df[0]$ is a finite-dimensional linear transformation of rank m, and f maps open sets of U onto relatively open sets of M.*

REMARK 4.5.6 Suppose that $F : \mathbb{F}^n \times \mathbb{F}^m \to \mathbb{F}^n$ is an $\mathbb{F}$-analytic mapping with $F(x_0, y_0) = z_0$ and that $\partial_x F[(x_0, y_0)]$ is a bijection on $\mathbb{F}^n$. Then the analytic implicit function theorem 4.5.4 defines an $\mathbb{F}$-analytic manifold of dimension m by the equation $F(x, y) = z_0$ for (x, y) in a neighbourhood of (x_0, y_0). □

THEOREM 4.5.7 (Composition of Analytic Functions) *Suppose X, Y and Z are Banach spaces and that $U \subset X$ and $V \subset Y$ are open. Suppose that $F : U \to V$ and $G : V \to Z$ are $\mathbb{F}$-analytic. Then $G \circ F : U \to Z$ is $\mathbb{F}$-analytic.*

Proof. Let $W = U \times V \times Z$ and define an analytic function $H : W \to Y \times Z$ by

$$H(x, y, z) = (F(x) - y, G(y) - z).$$

Let $x_0 \in U$, $y_0 = F(x_0) \in V$ and $z_0 = G(y_0)$. Then $H(x_0, y_0, z_0) = (0, 0) \in Y \times Z$ and

$$\partial_{(y,z)} H[(x_0, y_0, z_0)](y, z) = (-y, dG[y_0]y - z) = (\hat{y}, \hat{z})$$

if and only if

$$y = -\hat{y} \text{ and } z = -\hat{z} - dG[y_0]\hat{y}.$$

Thus $\partial_{(y,z)} H[(x_0, y_0, z_0)]$ is a homeomorphism. By the analytic implicit function theorem 4.5.4 the solution set, in a neighbourhood of (x_0, y_0, z_0), of the equation $H(x, y, z) = (0, 0)$ is described by $(y, z) = (\hat{Y}(x), \hat{Z}(x))$, where $(\hat{Y}, \hat{Z})$ denotes a $Y \times Z$-valued, $\mathbb{F}$-analytic function defined on a neighbourhood of x_0 in X. But $H(x, y, z) = 0$, $(x, y, z) \in W$ implies that $G(F(x)) = z$. Hence $G(F(x)) = \widehat{Z}(x)$ for x in a neighbourhood of $x_0 \in X$. Since $x_0 \in U$ was chosen arbitrarily, it is immediate that $G \circ F$ is analytic.   □

The following is the $\mathbb{F}$-analytic version of Proposition 3.6.1 (in the same notation).

PROPOSITION 4.5.8   (Analytic Perturbation of a Simple Eigenvalue)
Let X and Y be Banach spaces with X continuously embedded in Y and let $s \mapsto L(s)$ be $\mathbb{F}$-analytic from $(-1, 1)$ into $\mathcal{L}(X, Y)$. Suppose also that μ_0 is a simple eigenvalue of $L(0)$ with normalized eigenvector ξ_0. Then there exists $\epsilon > 0$ and an $\mathbb{F}$-analytic curve $\{(\mu(s), \xi(s)) : s \in (-\epsilon, \epsilon)\} \subset \mathbb{R} \times X$ such that $(\mu(0), \xi(0)) = (\mu_0, \xi_0)$,

$$L(s)\xi(s) = \mu(s)\iota\,\xi(s) \ \ and \ \ \xi(s) = \xi_0 + \eta(s),$$

where $\iota\eta(s) \in \text{range}\,(L(0) - \mu_0\iota)$. Moreover, $\mu(s)$ is a simple eigenvalue of $L(s)$ and if $|s| < \epsilon$ and μ is an eigenvalue of $L(s)$ with $|\mu_0 - \mu| < \epsilon$ then $\mu = \mu(s)$.

Proof. The proof is identical to that of Proposition 3.6.1, using the analytic implicit function theorem 4.5.4.   □

4.6 NOTES ON SOURCES

The theory of analytic functions is covered in books by Dieudonné [28], Federer [29] and Narasimhan [46]. The proofs of the inverse and implicit function theorems, and the demonstration that the composition of analytic functions is analytic, are based on an approach to the Weierstrass preparation theorem in Narasimhan [46]. The treatment of the Faà de Bruno formula is taken from Fraenkel [30].

PART 2

Analytic Varieties

Chapter Five

Analytic Functions on $\mathbb{F}^n$

We now present some fundamental results about $\mathbb{F}$-analytic functions in finite dimensions.

5.1 PRELIMINARIES

Unless otherwise stated $\mathbb{F}$ denotes $\mathbb{R}$ or $\mathbb{C}$. Consider $\mathbb{F}^n$ as an $\mathbb{F}$-linear space the points of which are described by coordinates using the standard basis of real vectors of the form $(0, \cdots, 1, \cdots 0) \in \mathbb{F}^n$. For $x = (x_1, \cdots, x_n) \in \mathbb{F}^n$ and $p = (p_1, \cdots, p_n) \in \mathbb{N}_0^n$ let

$$|x|^2 = \sum_{j=1}^n |x_j|^2,$$

$$|p| = \sum_{j=1}^n p_j, \quad x^p = x_1^{p_1} \cdots x_n^{p_n}, \quad p! = p_1! p_2! \cdots p_n!$$

and

$$\frac{\partial^p f}{\partial x^p} = \frac{\partial f^{|p|}}{\partial x_1^{p_1} \partial x_2^{p_2} \cdots \partial x_n^{p_n}}.$$

For $n \in \mathbb{N}$, let $U \subset \mathbb{F}^n$ be an open neighbourhood of $0 \in \mathbb{F}^n$ and $f : U \to \mathbb{F}$ an $\mathbb{F}$-analytic function. Then an induction argument starting with the case $n = 2$ in (4.19) and (4.20) leads to

$$f(x) = \sum_{p \in \mathbb{N}_0^n} f_p x^p$$

where

$$f_p = \frac{1}{p!} \frac{\partial^p f}{\partial x^p}(0) \in \mathbb{F} \text{ and } \sup_{p \in \mathbb{N}_0^n} \frac{p!}{|p|!} |f_p| r^{|p|} < \infty$$

for some $r > 0$. This in turn leads to

$$\sum_{p \in \mathbb{N}_0^n} r^{|p|} |f_p| < \infty \text{ for some } r > 0.$$

A function so defined is analytic on the open neighbourhood $\left(B_r(\mathbb{F})\right)^n$ of 0 in $\mathbb{F}^n$. (Here $B_r(\mathbb{F})$ is the open disc with radius r centred at 0 in $\mathbb{F}$.)

DEFINITION 5.1.1 *If $U \subset \mathbb{C}^n$ is open and $f : U \to \mathbb{C}$ is $\mathbb{C}$-analytic with the property that $f(x) \in \mathbb{R}$ for all $x \in U \cap \mathbb{R}^n$ we say that f is real-on-real. Equivalently f is real-on-real if and only if for all $x_0 \in U \cap \mathbb{R}^n$ all the terms f_p in the expansion of f at x_0 are real-on-real multilinear forms on $\mathbb{C}^n$.*

We use "real-on-real", instead of "real", to emphasis the complex setting.

REMARK 5.1.2   Suppose that $f : \mathbb{C}^n \to \mathbb{C}$ is real-on-real. If the basis of $\mathbb{C}^n$ is changed to another real basis (a basis of vectors each of which has real coordinates with respect to the standard basis) and points of $\mathbb{C}^n$ are now described by coordinates $(\zeta_1, \cdots, \zeta_n)$ with respect to that basis, then the function f is a real-on-real function of $(\zeta_1, \cdots, \zeta_n) \in \mathbb{C}^n$. Therefore a real-on-real $\mathbb{C}$-analytic function remains real-on-real after a real coordinate change of the independent variables.   □

Now we introduce linear spaces of $\mathbb{F}$-analytic functions as follows. For $q \in \mathbb{N}$ and $r > 0$, let $\mathcal{B}_r^q = \left(B_{r^{q+1}}(\mathbb{F})\right)^{n-1} \times B_r(\mathbb{F})$, an open neighbourhood of 0 in $\mathbb{F}^n$, and let C_r^q denote the space of $\mathbb{F}$-valued $\mathbb{F}$-analytic functions u on $\mathcal{B}_r^q$ with $u(0) = 0$ of the form

$$u(x) = \sum_{p \in \mathbb{N}_0^n,\, p \neq 0} u_p\, x^p$$

where

$$\sum_{p \in \mathbb{N}_0^n,\, p \neq 0} |u_p| r^{(q+1)|p| - q p_n} = \|u\|_{r,q} < \infty.$$

PROPOSITION 5.1.3 $\left(C_r^q, \|\cdot\|_{r,q}\right)$ *is a Banach space. In fact it is a Banach algebra since it is closed under multiplication and $\|uv\|_{r,q} \leq \|u\|_{r,q} \|v\|_{r,q}$.*

Proof.   Suppose that $\{u^k\}_{k \in \mathbb{N}} \subset C_r^q$ is a Cauchy sequence. Then the corresponding sequence $\{u_p^k\}_{k \in \mathbb{N}}$, $p \neq 0$, of Taylor coefficients of u at 0 is Cauchy in $\mathbb{F}$. Let $u_p^k \to u_p$ in $\mathbb{F}$. Since Cauchy sequences are bounded there exists a constant M such that for all P, $k \in \mathbb{N}$,

$$\sum_{p \in \mathbb{N}_0^n,\, 0 < |p| \leq P} |u_p^k| r^{(q+1)|p| - q p_n} \leq M.$$

In the limit as $k \to \infty$ with P fixed this yields

$$\sum_{p \in \mathbb{N}_0^n,\, 0 < |p| \leq P} |u_p| r^{(q+1)|p| - q p_n} \leq M.$$

Since this is true for all $P \in \mathbb{N}$

$$\sum_{p \in \mathbb{N}_0^n,\, p \neq 0} |u_p| r^{(q+1)|p| - q p_n} < \infty.$$

Let u denote the function in C_r^q the Taylor coefficients of which are u_p, $p \neq 0$, and $u_0 = 0$. Let $\epsilon > 0$ be given and, using the fact that $\{u^k\}$ is Cauchy in C_r^q, choose $N \in \mathbb{N}$ such that for all $P \in \mathbb{N}$,

$$\sum_{\substack{p \in \mathbb{N}_0^n, p \neq 0 \\ |p| \leq P}} |u_p^n - u_p^k| r^{(q+1)|p| - qp_n} \leq \epsilon$$

if k, $n \geq N$. With P arbitrary (but fixed) let $n \to \infty$ to obtain

$$\sum_{\substack{p \in \mathbb{N}_0^n, p \neq 0 \\ |p| \leq P}} |u_p - u_p^k| r^{(q+1)|p| - qp_n} \leq \epsilon.$$

Since this is true for any P, it follows that $\|u_k - u\|_{r,q} \leq \epsilon$ for all $k \geq N$. Thus C_r^q is a Banach space.

Now to prove that it is an algebra, let $\{u_p\}$ and $\{v_p\}$, $p \neq 0$, be the Taylor coefficients of u, $v \in C_r^q$ where $u_0 = v_0 = 0$. Let w be the product of u and v on the open set $\mathcal{B}_r^q$. Then, by Cauchy's product formula,

$$w_p = \sum_{\substack{s, t \in \mathbb{N}_0^n \\ s, t \neq 0 \\ s + t = p}} u_s v_t.$$

Therefore

$$\sum_{p \in \mathbb{N}_0^n, p \neq 0} |w_p| r^{(q+1)|p| - qp_n}$$

$$\leq \sum_{p \in \mathbb{N}_0^n, p \neq 0} \sum_{\substack{s, t \in \mathbb{N}_0^n \\ s, t \neq 0 \\ s + t = p}} \left(r^{(q+1)|s| - qs_n} |u_s| \right) \left(r^{(q+1)|t| - qt_n} |v_t| \right)$$

$$= \left(\sum_{s \in \mathbb{N}_0^n, s \neq 0} |u_s| r^{(q+1)|s| - qs_n} \right) \times \left(\sum_{t \in \mathbb{N}_0^n, t \neq 0} |v_p| r^{(q+1)|t| - qt_n} \right)$$

$$= \|u\|_{r,q} \|v\|_{r,q}.$$

This proves that $w \in C_r^q$ and $\|w\|_{r,q} \leq \|u\|_{r,q} \|v\|_{r,q}$ when $w = uv$, and the proof is complete. $\qquad \square$

Now we define linear operators A, L and B on C_r^q by

$$Au(x) = \sum_{\substack{p \in \mathbb{N}_0^n, \\ p_n < q}} u_p\, x^p,$$

$$Lu(x) = \sum_{\substack{p \in \mathbb{N}_0^n, \\ p_n \geq q}} u_p\, x_1^{p_1} \cdots x_{n-1}^{p_{n-1}} x_n^{p_n - q} = x_n^{-q}(I - A)u(x),$$

$$Bu(x) = \sum_{\substack{p \in \mathbb{N}_0^n, \\ p_1 + \cdots + p_{n-1} > 0,}} u_p x^p, \qquad x \in \mathcal{B}_r^q, \quad u \in C_r^q.$$

LEMMA 5.1.4

$A \in \mathcal{L}(C_r^q, C_r^q)$ *with* $\|A\| = 1$;

$L \in \mathcal{L}(C_r^q, C_r^q)$ *with* $\|L\| \leq r^{-q}$;

$\|Bu\|_{r,q} \leq C(u)r^{1+q}$, *where* $C(u)$ *is a constant determined by* u.

Proof. A and B are projections and the results about them are obvious from their definitions and that of the norm on C_r^q. Note that for $p \in \mathbb{N}_0^n$, $p \neq 0$, the coefficient $(Lu)_p$ coincides with $u_{(p_1,\cdots,p_{n-1},q+p_n)}$ and hence

$$\|Lu\|_{r,q} = \sum_{p \in \mathbb{N}_0^n, p \neq 0} |(Lu)_p|\, r^{(q+1)(p_1 + \cdots + p_{n-1}) + p_n}$$

$$= \sum_{p \in \mathbb{N}_0^n, p \neq 0} |u_{(p_1,\cdots p_{n-1},q+p_n)}|\, r^{(q+1)(p_1 + \cdots + p_{n-1}) + p_n}$$

$$= r^{-q} \sum_{p \in \mathbb{N}_0^n, p \neq 0} |u_{(p_1,\cdots p_{n-1},q+p_n)}|\, r^{(q+1)(p_1 + \cdots + p_{n-1}) + q + p_n}$$

$$\leq r^{-q} \|u\|_{r,q}.$$

Therefore $L \in \mathcal{L}(C_r^q, C_r^q)$ and $\|L\| \leq r^{-q}$. $\qquad\qquad\square$

5.2 WEIERSTRASS DIVISION THEOREM

THEOREM 5.2.1 *Suppose* $0 \in U$ *(open)* $\subset \mathbb{F}^n$, $f : U \to \mathbb{F}$ *is analytic,* $f(0) = 0$ *and, for* $(0, \cdots, 0, x_n) \in U$,

$$f(0, \cdots, 0, x_n) = x_n^q v(x_n) \text{ where } v(0) \neq 0 \text{ and } q \geq 1.$$

Let $g : U \to \mathbb{F}$ *be any* $\mathbb{F}$-*analytic function with* $g(0) = 0$.
Then for some $r > 0$,

$$g(x_1, \cdots, x_n)$$

$$= h(x_1, \cdots, x_n)f(x_1, \cdots, x_n) + \sum_{k=0}^{q-1} h_k(x_1, \cdots, x_{n-1})x_n^k$$

for all $(x_1, \cdots, x_n) \in U_0 = \mathcal{B}_r^q$, *where h is analytic on U_0 and h_k is analytic on* $V = \left(B_{r^{q+1}}(\mathbb{F})\right)^{n-1}$. *The functions h_k and h are uniquely determined by f and g. If $\mathbb{F}^n = \mathbb{C}^n$ and f and g are real-on-real, then h_k and h are real-on-real.*

Proof. Without loss of generality suppose that $v(0) = 1$. Choose $\hat{r}$ such that f, $g \in C_r^q$ for all r with $0 < r \leq \hat{r}$, where q is given by the hypothesis on f. We will prove the theorem by showing that the linear operator $\Gamma : C_r^q \to C_r^q$ defined, in the notation of Lemma 5.1.4, by

$$\Gamma u(x) = f(x) L u(x) + A u(x), \quad x \in \mathcal{B}_r^q,$$

is a bijection on C_r^q provided $r > 0$ is sufficiently small. Let $v_0(x) = x_n^q$, $x \in U$. Then $\|v_0\|_{r,q} = r^q$,

$$f(0, \cdots, 0, x_n) = v(x_n) v_0(0, \cdots, x_n) \text{ and hence } f - v v_0 = Bf,$$

where v is in the statement of the theorem. Therefore by Proposition 5.1.3 and Lemma 5.1.4,

$$\begin{aligned} \|f - v_0\|_{r,q} &\leq \|f - v v_0\|_{r,q} + \|v_0(1 - v)\|_{r,q} \\ &\leq \|Bf\|_{r,q} + \|v_0\|_{r,q}\|1 - v\|_{r,q} \\ &\leq C(f) r^{1+q} + r^q \|1 - v\|_{r,q}. \end{aligned}$$

From the definition of L, $(\Gamma - I)u = (f - v_0)Lu$ for $u \in C_r^q$, and hence

$$\begin{aligned} \|(\Gamma - I)u\|_{r,q} &= \|(f - v_0)Lu\|_{r,q} \leq \|Lu\|_{r,q}\|(f - v_0)\|_{r,q} \\ &\leq r^{-q}\|u\|_{r,q}\left(C(f)r^{1+q} + r^q\|1 - v\|_{r,q}\right). \end{aligned}$$

Since $v(0) = 1$, $\|1 - v\|_{r,q} \to 0$ as $r \to 0$, and it follows that $\Gamma - I \in \mathcal{L}(C_r^q, C_r^q)$ with norm less that 1 if $r > 0$ is sufficiently small. Hence Γ is a bijection on C_r^q. Hence for $g \in C_r^q$ there is a unique $u \in C_r^q$ with $\Gamma u = g$. The uniqueness of h and h_k now follow from the definition of L and A.

If $\mathbb{F} = \mathbb{C}$ and f and g are real-on-real, then the theorem restricted to $\mathbb{R}^n \cap U$ yields $\mathbb{R}$-analytic functions h and h_k. By uniqueness when $\mathbb{F} = \mathbb{C}$ it follows that h and h_k are real-on-real. The proof is complete. $\qquad\square$

REMARK 5.2.2 The division theorem can be interpreted as saying that f divides g with a remainder that is a polynomial of degree at most $(q - 1)$ in x_n, with coefficients that are analytic functions of $(x_1, \cdots, x_{n-1})$. $\qquad\square$

5.3 WEIERSTRASS PREPARATION THEOREM

The next result is a special case of the division theorem.

THEOREM 5.3.1 *Suppose* $0 \in U$ *(open)* $\subset \mathbb{F}^n$, $f : U \to \mathbb{F}$ *is analytic,* $f(0) = 0$ *and, for* $(0, \cdots, 0, x_n) \in U$,

$$f(0, \cdots, 0, x_n) = x_n^q v(x_n) \text{ where } v(0) \neq 0 \text{ and } q \geq 1.$$

Then for $r > 0$ *sufficiently small,*

$$h(x_1, \cdots, x_n) f(x_1, \cdots, x_n) = x_n^q + \sum_{k=0}^{q-1} a_k(x_1, \cdots, x_{n-1}) x_n^k,$$

for all $(x_1, \cdots, x_n) \in U_0 = B_r^q$, *where* $h(0) \neq 0$, h *and* $1/h$ *are analytic on* U_0, *and* a_k *is analytic on* $V = \left(B_{r^{q+1}}(\mathbb{F})\right)^{n-1}$ *with* $a_k(0) = 0$. *The functions* a_k *and* h *are uniquely determined by* f. *If* $U \subset \mathbb{C}^n$ *and* f *is real-on-real, then* h *and* a_k *are real-on-real.*

Proof. In the Weierstrass division theorem 5.2.1, choose $g(x) = x_n^q$. Then

$$x_n^q - \sum_{k=0}^{q-1} h_k(x_1, \cdots, x_{n-1}) x_n^k = h(x_1, \cdots, x_n) f(x_1, \cdots, x_n)$$

in the neighbourhood U_0 of $0 \in \mathbb{F}^n$. In particular,

$$x_n^q - \sum_{k=0}^{q-1} h_k(0, \cdots, 0) x_n^k = h(0, \cdots, 0, x_n) v(x_n) x_n^q.$$

Thus $h_k(0, \cdots, 0) = 0$, $0 \leq k \leq q-1$ and $h(0, \cdots, 0) \neq 0$, since $v(0) \neq 0$. Since the composition of analytic functions is analytic, $1/h$ is analytic in a neighbourhood of 0. Let $a_k = -h_k$ to complete the proof. $\square$

REMARK 5.3.2 The preparation theorem has a corollary that in a neighbourhood of $0 \in \mathbb{F}^n$ the solution set of $f(x_1, \cdots, x_n) = 0$ coincides with the zero set of a polynomial in x_n (the coefficients of which are analytic functions on V) of the form

$$x_n^q + \sum_{k=0}^{q-1} a_k(x_1, \cdots, x_{n-1}) x_n^k \text{ where } a_k(0, \cdots, 0) = 0. \tag{5.1}$$

The question of how the roots x_n depend on $(x_1, \cdots, x_{n-1})$ is the basis of the theory to follow. $\square$

5.4 RIEMANN EXTENSION THEOREM

We begin our study of the level sets of $\mathbb{F}$-analytic functions with a general observation on metric spaces.

LEMMA 5.4.1   *(a) Suppose that a metric space U has a subset G_1 that is dense and connected. If $G_1 \subset G \subset U$, then G is connected.*

(b) Suppose that G is a closed subset of a non-empty connected metric space U such that $U \setminus G$ is dense in U. Suppose also that each $x \in G$ has an open neighbourhood V_x such that $V_x \setminus G$ is connected. Then $U \setminus G$ is connected.

Proof. (a) Suppose that G is not connected. Then there exist non-empty, closed subsets F_1 and F_2 of U such that

$$G \subset F_1 \cup F_2, \quad F_1 \cap F_2 \cap G = \emptyset \text{ and } G \cap F_i \neq \emptyset, \quad i = 1, 2.$$

Since G_1 is connected, G_1 is a subset of one of them, say $G_1 \subset F_1$. Hence $U = \overline{G_1} \subset F_1$, which is a contradiction. This proves (a).

(b) Suppose that $U \setminus G$ is not connected. Then there exist non-empty, open, disjoint sets O_1 and O_2 whose union is $U \setminus G$. Let $x \in G$. Since $V_x \setminus G$ is connected, it is a subset of O_1 or of O_2, but not of both.

Let $G_i = \{x \in G : V_x \setminus G \subset O_i\}$, $i \in \{1, 2\}$, so that $G = G_1 \cup G_2$. Suppose that $x_k \in G_i$ (i fixed) and $x_k \to x \in G_j$, $j \in \{1, 2\}$ (recall that G is closed). Then $V_{x_k} \setminus G$ is a subset of O_1 or of O_2. Since $U \setminus G$ is dense, $(V_x \setminus G) \cap (V_{x_k} \setminus G) \neq \emptyset$ for all k sufficiently large. Hence $V_{x_k} \setminus G \subset O_j$ for k sufficiently large. Therefore $j = i$ and G_i is closed, $i \in \{1, 2\}$.

Now let $U_i = O_i \cup G_i$. Clearly U_i, $i \in \{1, 2\}$ are non-empty disjoint subsets of U whose union is U. Suppose that $\{x_k\}$ is a convergent sequence in U_i, with limit x. If, for infinitely many k, $x_k \in G_i$, then $x \in G_i \subset U_i$. Suppose instead that $x_k \in O_i$ and $x \notin O_i$. Then $x \in G$ and $x_k \in V_x \setminus G$ for all k sufficiently large. Therefore $V_x \setminus G \subset O_i$. Hence $x \in G_i$. This shows that U_i is closed in U, $i \in \{1, 2\}$, and hence U is not connected, a contradiction that proves (b).   □

PROPOSITION 5.4.2   *Suppose that $U \subset \mathbb{F}^n$ is open, connected and $g_k : U \to \mathbb{F}$ is $\mathbb{F}$-analytic, $1 \leq k \leq m$. Let $E = \{x \in U : g_k(x) = 0 \in \mathbb{F}, 1 \leq k \leq m\}$.*
(a) If $E \neq U$, then $U \setminus E$ is dense in U.
(b) If, in addition, $\mathbb{F} = \mathbb{C}$, then $U \setminus E$ is connected.

Proof. (a) If $U \setminus E$ is not dense then E contains an open subset of U on which all the functions g_k are zero. Hence they are all identically zero on U, by Lemma 4.3.9. Since $E \neq U$, this is a contradiction which proves (a).

(b) First we observe from Lemma 5.4.1(a) and part (a) above that it suffices to treat the case $m = 1$. Let $E = \{x \in U : g(x) = 0\}$ where $U \subset \mathbb{C}^n$ is open and $g : U \to \mathbb{C}$ is $\mathbb{C}$-analytic. By Lemma 5.4.1 (b) it will suffice, for every $x \in E$, to find an open neighbourhood $V_x \subset U$ such that $V_x \setminus E$ is connected. Without loss of generality suppose that $x = 0 \in U$ and that $g(0) = 0$. Suppose moreover that $g \not\equiv 0$ on U. Then we can choose the coordinates $(x_1, \cdots, x_n)$ such that $g(0, \cdots, 0, x_n) \not\equiv 0$. Since $x_n \mapsto g(0, \cdots, 0, x_n)$ is a $\mathbb{C}$-analytic function on a neighbourhood of $0 \in \mathbb{C}$, its zeros are isolated and therefore there exists $\epsilon > 0$ such that $g(0, \cdots, 0, x_n) \neq 0$ if $0 < |x_n| \leq \epsilon$. Hence, by continuity, there exists

$\delta > 0$ such that $g(x_1, \cdots, x_n) \neq 0$ if $\sum_{k=1}^{n-1} |x_k| < \delta$ and $\frac{1}{2}\epsilon < |x_n| < \epsilon$. Let

$$V_0 = \left\{(x_1, \cdots, x_n) : \sum_{k=1}^{n-1} |x_k| < \delta, \ |x_n| < \epsilon\right\}.$$

Now in $\mathbb{C}^n$ (but not in $\mathbb{R}^n$) the set

$$\tilde{V}_0 = \left\{(x_1, \cdots, x_n) : \sum_{k=1}^{n-1} |x_k| < \delta, \ \tfrac{1}{2}\epsilon < |x_n| < \epsilon\right\}$$

is a path-connected subset of V_0 on which g is nowhere zero. Moreover, for each fixed $(\hat{x}_1, \cdots, \hat{x}_{n-1})$ with $\sum_{k=1}^{n-1} |\hat{x}_k| < \delta$, the analytic function $x_n \mapsto g(\hat{x}_1, \cdots, \hat{x}_{n-1}, x_n)$ has at most a finite number of zeros with $|x_n| \leq 3\epsilon/4$ and the set

$$\{x \in U : x = (\hat{x}_1, \cdots, \hat{x}_{n-1}, x_n), \ |x_n| \leq 3\epsilon/4, g(x) \neq 0\} \subset \mathbb{C}^n$$

is path-connected. Since $\tilde{V}_0$ is path-connected, $V_0 \setminus E$ is path-connected, and hence connected. Since V_0 is an open neighbourhood of 0, where 0 represents an arbitrary point of E, this completes the proof.   □

The following is a particular case of a classical theorem which holds more generally.

THEOREM 5.4.3 (Riemann Extension Theorem) *Suppose that $U \subset \mathbb{C}^n$ is open and $g_k : U \to \mathbb{C}$ is $\mathbb{C}$-analytic, $1 \leq k \leq m$. Let*

$$E = \{x \in U : g_k(x) = 0 \text{ for all } k, \ 1 \leq k \leq m\}$$

and suppose that f is $\mathbb{C}$-analytic on $U \setminus E$ with

$$\sup\{|f(x)| : x \in U \setminus E\} < \infty.$$

Then there exists a function $\tilde{f}$ which is $\mathbb{C}$-analytic on U and $f = \tilde{f}$ on $U \setminus E$.

Proof. Since

$$U \setminus E = \cup_{k=1}^{m} (U \setminus E_k) \text{ where } E_k = \{x \in U : g_k(x) = 0\}$$

it suffices to prove the required result when $m = 1$. Since analyticity is defined locally it will suffice to choose $x \in E$ and to show that the result is true when U is replaced by some open neighbourhood of x. Without loss of generality suppose that $0 \in E$ is the point in question. Let $\epsilon, \delta > 0$ and V_0, a neighbourhood of $0 \in U \cap E$, be as defined in the proof of part (b) of the last proposition. Then for any fixed $(x_1, \cdots, x_{n-1})$ with $\sum_{k=0}^{n-1} |x_k| < \delta$, the set

$$\{z \in \mathbb{C} : |z| \leq 3\epsilon/4 \text{ and } (x_1, \cdots, x_{n-1}, z) \in E\}$$
$$\subset \{z \in \mathbb{C} : |z| \leq 3\epsilon/4 \text{ and } g_1(x_1, \cdots, x_{n-1}, z) = 0\}$$

is finite and therefore $z \mapsto f(x_1, \cdots, x_{n-1}, z)$ is $\mathbb{C}$-analytic except at finitely many points and bounded on $\{z \in \mathbb{C} : |z| \leq 3\epsilon/4\}$. Note also that $z \mapsto f(x_1, \cdots, x_{n-1}, z)$ is analytic in a neighbourhood of the circle $|z| = 3\epsilon/4$ for any such fixed $(x_1, \cdots, x_{n-1})$, since then $(x_1, \cdots, x_{n-1}, z) \in \tilde{V}_0$ and g does not vanish on $\tilde{V}_0$. Therefore the singularities of this function of z must be removable and, by Cauchy's integral formula, the function

$$\tilde{f}(x_1, \cdots, x_n) = \frac{1}{2\pi} \int_{-\pi}^{\pi} \frac{f(x_1, \cdots, x_{n-1}, 3\epsilon e^{it}/4)}{3\epsilon e^{it}/4 - x_n} \left(\frac{3\epsilon e^{it}}{4} \right) dt \qquad (5.2)$$

extends f to all of V_0. For fixed $(x_1, \cdots, x_{n-1})$ as above, $x_n \mapsto \tilde{f}(x_1, \cdots, x_n)$ is $\mathbb{C}$-analytic. However it is clear, from the definition of $\tilde{f}$ given in (5.2) that $\tilde{f}$ is $\mathbb{C}$-analytic on $V_0 \subset \mathbb{C}^n$. This completes the proof. $\qquad \square$

The function $\tilde{f}$ is called an analytic extension of f.

5.5 NOTES ON SOURCES

These classical results are in the books by Chow and Hale [19], Dieudonné [28], Federer [29], Golubitsky and Guillemin [34]. For results on functions of a complex variable, see Ahlfors [1], Cartan [15], Chatterji [17].

Chapter Six

Polynomials

6.1 CONSTANT COEFFICIENTS

A polynomial of the form

$$A(Z) = a_p Z^p + \ldots + a_1 Z + a_0, \; p \in \mathbb{N}_0, \; a_p \neq 0,$$

with complex coefficients is said to have degree p and a complex number z such that $A(z) = 0$ is called a root of $A(Z)$. The fundamental theorem of algebra says that $A(Z)$ has at most p roots, $z_1, \cdots, z_k$, say, and that

$$A(Z) = a_p(Z - z_1)^{m_1} \cdots (Z - z_k)^{m_k}, \text{ where } m_1 + \cdots + m_k = p \quad (6.1)$$

and the z_j s are distinct. In this factorization of $A(Z)$ over $\mathbb{C}$, the number m_j is called the multiplicity of the root z_j. If $m_j = 1$ then z_j is called a simple root of $A(Z)$, otherwise it is a multiple root. The coefficient of Z^p is called the principal coefficient of $A(Z)$.

Continuous Dependence of Roots

PROPOSITION 6.1.1 *Let $p \geq 1$. (a) If $\widehat{z} \in \mathbb{C}$ is a simple root of a polynomial*

$$Z^p + \widehat{a}_{p-1} Z^{p-1} + \cdots + \widehat{a}_0$$

with complex coefficients, then there is a $\mathbb{C}$-analytic function f, defined in a neighbourhood of $(\widehat{a}_0, \cdots, \widehat{a}_{p-1})$, such that $z = f(a_0, \cdots, a_{p-1})$ is a simple root of the polynomial $Z^p + a_{p-1}Z^{p-1} + \cdots + a_0$ and $\widehat{z} = f(\widehat{a}_0, \cdots, \widehat{a}_{p-1})$. If, in addition, $\widehat{a}_0, \cdots, \widehat{a}_{p-1}, \widehat{z}, a_0, \cdots, a_{p-1} \in \mathbb{R}$, then $f(a_0, \cdots, a_{p-1}) \in \mathbb{R}$.
(b) Suppose that $\widehat{z} \in \mathbb{C}$ is a root of multiplicity $q \geq 1$ of the polynomial in part (a), and the distance of all the other roots from $\widehat{z}$ is at least $\widehat{\epsilon} > 0$. Then, for all ϵ with $0 < \epsilon < \widehat{\epsilon}$ there exists $\delta > 0$ such that the polynomial

$$Z^p + a_{p-1}Z^{p-1} + \ldots + a_0$$

has exactly q complex roots, counted according to their multiplicities, in the set

$$\{z \in \mathbb{C} : |z - \widehat{z}| < \epsilon\}$$

provided that $|a_0 - \widehat{a}_0|, \ldots, |a_{p-1} - \widehat{a}_{p-1}| < \delta$.
(c) For all $\epsilon > 0$, there exist $\delta > 0$ such that $|z| < \epsilon$ for all $z \in \mathbb{C}$ with

$$z^p + a_{p-1}z^{p-1} + \cdots + a_0 = 0 \text{ when } |a_0|, \cdots, |a_{p-1}| < \delta.$$

Similarly, if $|a_0| + \cdots |a_{p-1}| \leq M$ *then* $|z| \leq m(M)$, *where* $m(M)$ *depends only on* M.

Proof. (a) Since $\hat{z}$ is a simple root of the polynomial,

$$\frac{d}{dZ}(Z^p + \hat{a}_{p-1}Z^{p-1} + \ldots + \hat{a}_0)\bigg|_{Z=\hat{z}} \neq 0.$$

Therefore the analytic implicit function theorem 4.5.4 ensures the existence of a $\mathbb{C}$-analytic function f such that, for all $(a_0, \ldots, a_{p-1})$ close to $(\hat{a}_0, \ldots, \hat{a}_{p-1})$ in $\mathbb{C}^p$,

$$\{Z^p + a_{p-1}Z^{p-1} + \cdots + a_0\}\big|_{Z=f(a_0, \cdots, a_{p-1})} \equiv 0$$

and

$$\frac{d}{dZ}(Z^p + a_{p-1}Z^{p-1} + \ldots + a_0)\bigg|_{Z=f(a_0, \ldots, a_{p-1})} \neq 0.$$

The same reasoning in the case of a polynomial with real coefficients completes the proof of the first part.

(b) Part (b) is immediate from Rouché's theorem. (c) If part (c) is false, then there exist sequences

$$\{a_{0,n}\}, \cdots, \{a_{p-1,n}\}, \{z_n\} \subset \mathbb{C}$$

such that

$$\lim_{n\to\infty} a_{j,n} = 0 \text{ for } 0 \leq j \leq p-1, \quad \lim_{n\to\infty} |z_n| \in (0, \infty]$$

and, for all $n \in \mathbb{N}$,

$$z_n^p + a_{p-1,n}z_n^{p-1} + \cdots + a_{0,n} = 0.$$

Thus we obtain the contradiction that

$$z_n = -a_{p-1,n} - \frac{a_{p-2,n}}{z_n} - \cdots - \frac{a_{0,n}}{z_n^{p-1}} \to 0,$$

and the proof is complete. $\square$

Greatest Common Divisors

Consider two polynomials of the complex variable Z, with constant coefficients in $\mathbb{C}$, given by

$$A(Z) = a_p Z^p + \cdots + a_1 Z + a_0, \quad p \geq 1, \quad a_p \neq 0,$$
$$B(Z) = b_q Z^q + \cdots + b_1 Z + b_0, \quad q \geq 1, \quad b_q \neq 0.$$

We will say that their greatest common divisor is the polynomial with largest degree $m \in \mathbb{N}_0$ of the form $Z^m + c_{m-1}Z^{m-1} + \cdots + c_0$, with coefficients $c_j \in \mathbb{C}$, which divides both $A(Z)$ and $B(Z)$. Then $A(Z)$ and $B(Z)$ are said to be co-prime if the constant polynomial 1 of degree 0 is their greatest common divisor. An elementary criterion says that $A(Z)$ and $B(Z)$ are not co-prime if and only if there exist two polynomials $P(Z)$ and $Q(Z)$ such that

$$A(Z)P(Z) + B(Z)Q(Z) = 0,$$
$$P(Z) \not\equiv 0, \ Q(Z) \not\equiv 0,$$
$$\deg(P(Z)) < q, \ \deg(Q(Z)) < p.$$

Equivalently, if $P(Z) = c_{q-1}Z^{q-1} + \cdots + c_1 Z + c_0$ and $Q(Z) = d_{p-1}Z^{p-1} + \cdots + d_1 Z + d_0, p \geq q$, the equation $\mathbf{A}x = \mathbf{0}$ has a solution

$$\mathbf{x} = (c_0, \cdots, c_{q-1}, d_0, \cdots, d_{p-1})^T \neq 0$$

if and only if $A(Z)$ and $B(Z)$ are not co-prime, where the square matrix $\mathbf{A}$ is given by

$$\mathbf{A_{ij}} = \begin{cases} a_{i-j}, & 0 \leq i - j \leq p, \ 1 \leq j \leq q \\ b_{q+i-j}, & 0 \leq j - i \leq q, \ q+1 \leq j \leq p+q \\ 0, & \text{otherwise} \end{cases}.$$

The complex $(p+q) \times (p+q)$ matrix $\mathbf{A}$ is called the resultant matrix of $A(Z)$ and $B(Z)$. Its determinant, $R(a_0, \cdots, a_p; b_0, \ldots, b_q)$, is called the resultant of $A(Z)$ and $B(Z)$. The resultant is a polynomial in the coefficients $a_0, \cdots, a_p, b_0, \cdots, b_q$ which is zero if and only if $A(Z)$ and $B(Z)$ are not co-prime.

LEMMA 6.1.2 *Let $A(Z)$ be a polynomial of degree $p \geq 1$ and let $A'(Z)$ denote its derivative. Denote by $C(Z)$ the greatest common divisor of $A(Z)$ and $A'(Z)$ and let $A(Z) = C(Z)E(Z)$ for some polynomial $E(Z)$. Then $E(Z)$ and $E'(Z)$ are co-prime and $E(Z)$ has the same roots as $A(Z)$.*

Proof. Suppose $A(Z)$ is given by (6.1). Then $A'(Z) = P(Z)C(Z)$ where

$$C(Z) = \prod_{\hat{k}=1}^{\hat{m}} (Z - \hat{z}_1)^{\hat{m}_1 - 1} \cdots (Z - \hat{z}_{\hat{k}})^{\hat{m}_{\hat{k}} - 1}$$

and $\{\hat{z}_{\hat{k}} : \hat{k} = 1, \cdots \hat{m}\}$, $\hat{z}_{\hat{k}}$ of multiplicity $\hat{m}_{\hat{k}} > 1$, denotes the set of multiple roots of $A(Z)$, and no root of $A(Z)$ is a root of $P(Z)$. Hence $C(Z)$ is the greatest common divisor of $A(Z)$ and $A'(Z)$, and $A(Z) = C(Z)E(Z)$ where all the roots of $E(Z)$ are simple and the roots of $E(Z)$ coincide with those of $A(Z)$. $\square$

For future reference, note that when $p = q$ the matrix $\mathbf{A}$ has the form

$$
\begin{bmatrix}
a_0 & 0 & \cdots & 0 & b_0 & 0 & \cdots & 0 \\
a_1 & a_0 & \cdots & 0 & b_1 & b_0 & \cdots & 0 \\
\vdots & \vdots & & \vdots & \vdots & \vdots & & \vdots \\
a_{p-1} & a_{p-2} & \cdots & a_0 & b_{p-1} & b_{p-2} & \cdots & b_0 \\
a_p & a_{p-1} & \cdots & a_1 & b_p & b_{p-1} & \cdots & b_1 \\
0 & a_p & \cdots & a_2 & 0 & b_p & \cdots & b_2 \\
\vdots & \vdots & & \vdots & \vdots & \vdots & & \vdots \\
0 & 0 & \cdots & a_p & 0 & 0 & \cdots & b_p
\end{bmatrix}
. \qquad (6.2)
$$

Discriminant of a Polynomial

Consider the polynomial

$$ A(Z) = a_p Z^p + \ldots + a_1 Z + a_0, \quad p \geq 1, \ a_p \neq 0, $$

the coefficients of which are complex. It is clear from (6.1) that $z_j \in \mathbb{C}$ is a multiple root of $A(Z)$ if and only if $(Z - z_j)$ is a common factor of $A(Z)$ and $A'(Z)$, where $A'(Z)$ denotes the polynomial obtained by differentiating A with respect to Z. Indeed $A(Z)$ has no multiple roots if and only if $A(Z)$ and $A'(Z)$ are co-prime. For $(a_0, \cdots, a_p) \in \mathbb{C}^{p+1}$ define $D(a_0, \ldots, a_p) \in \mathbb{C}$ to be the resultant of $A(Z)$ and $A'(Z)$. This is called the discriminant of $A(Z)$. Notice that D is a polynomial in the $p + 1$ variables $a_0, \ldots, a_p$ which vanishes exactly when $A(Z)$ has at least one multiple root.

EXAMPLE 6.1.3 (Quadratic polynomials)
    $p = 2$, $A(Z) = a_2 Z^2 + a_1 Z + a_0$, $A'(Z) = 2a_2 Z + a_1 = B(Z)$ and, with $a_2 \neq 0$,

$$
D(a_0, a_1, a_2) = R(a_0, a_1, a_2; b_0, b_1) =
\begin{vmatrix}
a_0 & b_0 & 0 \\
a_1 & b_1 & b_0 \\
a_2 & 0 & b_1
\end{vmatrix}
$$

$$
=
\begin{vmatrix}
a_0 & a_1 & 0 \\
a_1 & 2a_2 & a_1 \\
a_2 & 0 & 2a_2
\end{vmatrix}
= -a_2(a_1^2 - 4a_0 a_2),
$$

which vanishes exactly when the usual discriminant $a_1^2 - 4a_0 a_2 = 0$.   □

6.2 VARIABLE COEFFICIENTS

In the Weierstrass preparation theorem 5.3.1, we saw polynomials in x_n of the form

$$x_n^q + \sum_{k=0}^{q-1} a_k(x_1, \cdots, x_{n-1}) x_n^k,$$

in which the coefficients a_k are $\mathbb{F}$-analytic functions of $n - 1$ variables. In this section we consider such polynomials in the special case when $\mathbb{F} = \mathbb{C}$. Let $V \subset \mathbb{C}^m$, $m \geq 1$ be given by

$$V = \{(z_1, \cdots, z_m) \in \mathbb{C}^m : |z_j| < \delta,\ 1 \leq j \leq m\},$$

for some $\delta > 0$ and consider polynomials of the form

$$a_p(z_1, \cdots, z_m) Z^p + \cdots + a_1(z_1, \cdots, z_m) Z + a_0(z_1, \cdots, z_m),$$

where the coefficients $a_k : V \to \mathbb{C}$ are $\mathbb{C}$-analytic functions and $a_p \not\equiv 0$. (The case $m = 0$ will correspond to polynomials in Z with constant coefficients.) This polynomial, which we denote by $A(Z; z_1, \cdots, z_m)$, has degree p. The coefficients of the polynomial $B(Z; z_1, \cdots, z_m)$ are functions $b_k : V \to \mathbb{C}$, where it is understood that $A(Z; z_1, \cdots, z_m)$ and $B(Z; z_1, \cdots, z_m)$ may have different degrees. (When the meaning is clear we refer simply to polynomials A or B on V.) We say that a polynomial is real-on-real if and only if all its coefficients are real-on-real analytic functions. The discriminant $D = D(a_0, \cdots, a_p)$ of the polynomial A is an analytic function defined on $V \subset \mathbb{C}^m$ by

$$D(a_0, \cdots, a_p)(\xi) = D(a_0(\xi), \cdots, a_p(\xi)), \quad \xi = (z_1, \cdots, z_m) \in V. \qquad (6.3)$$

Greatest Common Divisors

From the previous viewpoint, $\{A(Z; z_1, \cdots, z_m) : (z_1, \cdots, z_m) \in V\}$ is a family of polynomials parameterized by $(z_1, \cdots, z_m) \in V$. The notion of the greatest common divisor of two polynomials A and B is therefore more subtle in this case.

DEFINITION 6.2.1 *Let $W \subset V$. The greatest common divisor of A and B on W is a polynomial $C(Z; z_1, \cdots, z_m)$ of degree d, say, where the coefficients c_k are $\mathbb{C}$-analytic on V, $c_d \equiv 1$ on V and, for every $(z_1, \cdots, z_m) \in W$, the polynomial $C(Z; z_1, \cdots, z_m)$ is the greatest common divisor of $A(Z; z_1, \cdots, z_m)$ and $B(Z; z_1, \cdots, z_m)$ in the sense of §6.1.*

The first question is whether in this general setting two polynomials have a greatest common divisor.

THEOREM 6.2.2 **(Euclid's Algorithm)** *Let polynomials*

$$A(Z; z_1, \cdots, z_m) \text{ and } B(Z; z_1, \cdots, z_m)$$

have degree p and q, respectively, where at least one of a_p and b_q is identically equal to 1 on V.

Then there exists a polynomial $C(Z; z_1, \cdots, z_m)$ of degree r on V and a $\mathbb{C}$-analytic function g, such that C is the greatest common divisor of A and B on the set $W = V \setminus G$, where $G = \{(z_1, \cdots, z_m) : g(z_1, \cdots, z_m) = 0\} \neq V$. Note that W is a dense connected subset of V, by Proposition 5.4.2.

Suppose that $a_p \equiv 1$. Then there is a polynomial $E(Z; z_1, \cdots, z_m)$ such that $A = CE$ for all $(Z; z_1, \cdots, z_m) \in \mathbb{C} \times V$. If $r = 0$ then $C \equiv 1$ on V.

Proof. Let

$$P_1 = A, \ m_1 = p \ \text{ and } \ P_2 = B, \ m_2 = q, \ \text{ if } p \geq q,$$
$$P_2 = A, \ m_2 = p \ \text{ and } \ P_1 = B, \ m_1 = q, \ \text{ if } p < q.$$

Suppose for $j \in \mathbb{N}$ that P_k, $1 \leq k \leq j+1$, are given polynomials of degree m_k such that the coefficient of Z^{m_k} is $g_k \neq 0$ and that m_k is a non-increasing function of k. Now let

$$Q(Z; z_1, \cdots, z_m) = Z^{m_j - m_{j+1}} g_j(z_1, \cdots, z_m) P_{j+1}(Z; z_1, \cdots, z_m)$$
$$- g_{j+1}(z_1, \cdots, z_m) P_j(Z; z_1, \cdots, z_m).$$

The crucial observation is that, on the set

$$W_j = \{(z_1, \cdots, z_m) \in V : g_{j+1}(z_1, \cdots, z_m) \neq 0\},$$

the greatest common divisor of P_j and P_{j+1} is also the greatest common divisor of Q and P_{j+1}. Obviously the degree of Q is smaller than the degree of P_j. If the degree of Q is not larger than that of P_{j+1} let $P_{j+2} = Q$; otherwise rename P_{j+1} as P_{j+2} and replace the old P_{j+1} by Q. This defines a set of polynomials $P_1, \cdots, P_{j+2}$ with $m_k > m_{k+2}$ for all $k \leq j$. Clearly this process terminates after a finite number of steps when

$$0 = g_{J+1} P_J - g_J Z^{m_J - m_{J+1}} P_{J+1}, \ \text{ say, for some } J \in \mathbb{N}.$$

Hence, at every point $(z_1, \cdots, z_m) \in V$ at which the product $g = g_1 \cdots g_{J+1}$ of the highest order coefficients of the P_js is non-zero, $\widehat{C} = P_{J+1}/g_{J+1}$ is the greatest common divisor of P_1 and P_2. This defines g and consequently G in the statement of the theorem. By definition, $g \neq 0$, and so $G \neq V$.

Now suppose that $a_p \equiv 1$. Then at every point $(z_1, \cdots, z_m) \in V \setminus G$ the roots of $\widehat{C}(Z; z_1, \cdots, z_m)$ form a subset of those of $A(Z; z_1, \cdots, z_m)$ and are therefore bounded when $(z_1, \cdots, z_m)$ lies in a compact subset of V, by Proposition 6.1.1 (c). Since the coefficients of $\widehat{C}$ are polynomial functions of the roots of $\widehat{C}$, it follows that these coefficients are bounded on subsets of $V \setminus G$ that are compact in V. Therefore, by the Riemann extension theorem 5.4.3, they have analytic extension to all of V. We denote by C the polynomial with the coefficients of $\widehat{C}$ extended to V. To obtain the existence of E, note first that a polynomial $\widehat{E}$ is defined on $V \setminus G$ by writing $A = C\widehat{E}$ on $V \setminus G$. The argument for extending $\widehat{E}$ as a polynomial on V is the same as that for $\widehat{C}$. $\qquad\square$

REMARKS 6.2.3 If A and B have coefficients that are real-on-real, then Euclid's algorithm obviously leads to a greatest common divisor C and a polynomial E, both of which are real-on-real.

If $a_p \equiv 1$, $b_q \equiv 1$, and $R(a_0, \cdots, a_{p-1}, 1; b_0, \cdots, b_{q-1}, 1) \not\equiv 0$ (in which case it is non-zero on a connected open dense set in V), we may take

$$g = R(a_0, \cdots, a_{p-1}, 1; b_0, \cdots, b_{q-1}, 1), \ r = 0 \ \text{and} \ c_0 = C \equiv 1.$$

If $r \geq 1$, then $R(a_0, \cdots, a_{p-1}, 1; b_0, \cdots, b_{q-1}, 1) \equiv 0$. $\qquad\square$

THEOREM 6.2.4 **(Simplifying a Polynomial)** *Let $A(Z; z_1, \cdots, z_m)$ be a polynomial of degree p with discriminant $D(a_0, \cdots, a_{p-1}, 1) \equiv 0$ and $a_p \equiv 1$ on V. Then there exists another polynomial $E(Z; z_1, \cdots, z_m)$ of degree q, say, with $e_q \equiv 1$ such that $E(Z; z_1, \cdots, z_m)$ has the same roots as $A(Z; z_1, \cdots, z_m)$, possibly with smaller multiplicities, and $D(e_0, \cdots, e_{m-1}, 1) \not\equiv 0$ on V. (In particular, for $(z_1, \cdots, z_m)$ in an open dense connected subset W of V, $E(Z; z_1, \cdots, z_m)$ has no multiple roots.) If A is real-on-real, then so is E. (E is called the simplification of A.)*

Proof. For $(z_1, \cdots, z_m) \in V$, the polynomial $A(Z; z_1, \cdots, z_m)$ has a multiple root if and only if its discriminant is zero. It therefore suffices to let E be the polynomial given by Euclid's algorithm for which $A = CE$ where C is the greatest common divisor of A and A' on W, an open dense connected subset of V. An appeal to Lemma 6.1.2 completes the proof. $\qquad\square$

THEOREM 6.2.5 **(Projection Lemma)** *Let $A_1(Z; z_1, \cdots, z_m)$ be a polynomial of degree p with $a_p \equiv 1$ on V and let $A_j(Z; z_1, \cdots, z_m)$ be a polynomial of degree at most $p - 1$, $2 \leq j \leq k$. Let*

$$\mathcal{A} = \{(z_1, \cdots, z_m) \in V : \text{there exists } z \in \mathbb{C}$$
$$\text{with } A_j(z; z_1, \cdots, z_m) = 0 \text{ for all } j \in \{1, \cdots, k\}\}.$$

Then there exists a finite family $\{R_\alpha : \alpha \in \Sigma\}$ of analytic functions on V such that

$$\mathcal{A} = \{(z_1, \cdots, z_m) \in V : R_\alpha(z_1, \cdots, z_m) = 0 \text{ for all } \alpha \in \Sigma\}.$$

If the polynomials A_j are real-on-real, then the functions in $\mathcal{A}$ are real-on-real.

REMARK 6.2.6 It is worth noting that this result is false in the setting of real-analytic functions. For example let $m = 1$ and $V = (-\epsilon, \epsilon)$, $k = 2$, $A_1(X, x_1) = X^2 - x_1$ and $A_2 \equiv 0$. Then $p = 2$ and $\mathcal{A} = [0, \epsilon)$ which is not the zero-set of any real-analytic function defined on $(-\epsilon, \epsilon)$, because of Lemma 4.3.9. $\qquad\square$

Proof. For any $t_2, \cdots, t_k \in \mathbb{C}$, let $R(t_2, \cdots, t_k; z_1, \cdots, z_m)$ denote the resultant of two polynomials, A_1 and $A_1 + \sum_{j=2}^k t_j A_j$, of the same degree. Let the coefficients of A_j be denoted by a_i^j, $1 \leq j \leq k$, and in (6.2) let $a_i = a_i^1$ and $b_i = a_i + \sum_{j=2}^k t_j a_i^j$. The resultant R may now be obtained from formula (4.2) for the determinant of the matrix (6.2) with these coefficients. By subtracting the jth column from the $p+j^{th}$ column, the coefficients a_i can be eliminated from the right half of the matrix without changing its determinant and the non-zero entries in the right half of the resulting matrix are all of the form $\sum_{j=2}^{k} t_j a_i^j$. Therefore

$$R(t_2, \cdots, t_k; z_1, \cdots, z_m) = \sum_{\substack{\alpha \in \mathbb{N}_0^{k-1} \\ |\alpha|=p}} t^\alpha R_\alpha(z_1, \cdots, z_m)$$

where $t = (t_2 \cdots, t_k)$ and $R_\alpha : V \to \mathbb{C}$ is analytic and independent of t.

Suppose that $(z_1, \cdots, z_m) \in \mathcal{A}$. Then, for some $z \in \mathbb{C}$,

$$A_j(z, z_1, \cdots, z_m) = 0 \text{ for all } j \in \{1, \cdots, k\}.$$

Therefore, for all $t_2, \cdots, t_k$, z is a common root of the polynomials A_1 and $A_1 + \sum_{j=2}^{k} t_j A_j$. Therefore $R(t_2, \cdots, t_k; z_1, \cdots, z_m) = 0$ for all $(t_2, \cdots, t_k) \in \mathbb{C}^{m-1}$ and so $R_\alpha(z_1, \cdots, z_m) = 0$ for all $\alpha \in \Sigma$ where $\Sigma = \{\alpha \in \mathbb{N}_0^{k-1} : |\alpha| = p\}$.

Conversely, suppose $(z_1, \cdots, z_m) \in V$ and $R_\alpha(z_1, \cdots, z_m) = 0$ for $\alpha \in \Sigma$. It follows that $R(t_2, \cdots, t_k; z_1, \cdots, z_m) = 0$ for $(t_2, \cdots, t_k) \in \mathbb{C}^{k-1}$. Therefore the polynomials A_1 and $A_1 + \sum_{j=2}^{k} t_j A_j$ have a common root (possibly depending on $(t_2, \cdots t_k)$) for all $(t_2, \cdots, t_k) \in \mathbb{C}^{k-1}$. Let $\zeta_1, \cdots, \zeta_\nu, \nu \leq p$, be the distinct roots of $A_1(Z; z_1, \cdots, z_m)$ and for $1 \leq i \leq \nu$ let

$$Y_i = \{(t_2, \cdots, t_k) \in \mathbb{C}^{k-1} : \sum_{j=2}^{k} t_j A_j(\zeta_i; z_1, \cdots, z_m) = 0\}.$$

Each Y_i is a linear subspace of $\mathbb{C}^{k-1}$ and their union is $\mathbb{C}^{k-1}$. Hence $Y_{i_0} = \mathbb{C}^{k-1}$ for some $i_0 \in \{1, \cdots, \nu\}$ and so $A_j(\zeta_{i_0}; z_1, \cdots, z_m) = 0$ for all $j \in \{1, \cdots, k\}$.

It is clear from the construction that if the polynomials A_j are real-on-real, then the functions R_α are real-on-real. This completes the proof. $\square$

6.3 NOTES ON SOURCES

The theory of polynomials with constant coefficients used here is extensively treated in Van der Waerden's classic [65]. The non-constant coefficient theory is to be found in Mumford [45] and Narasimhan [46].

Chapter Seven

Analytic Varieties

Once again the field $\mathbb{F}$ is either $\mathbb{C}$ or $\mathbb{R}$. Let $a \in \mathbb{F}^n$, $n \in \mathbb{N}$. Two subsets S and T of $\mathbb{F}^n$ are said to be equivalent at a if there is an open neighbourhood O of a such that $O \cap S = O \cap T$. We note that this is an equivalence relation on $2^{\mathbb{F}^n}$ and write $S \sim_a T$.

The corresponding equivalence class, denoted by $\gamma_a(S)$ for $S \subset \mathbb{F}^n$, is called the germ of S at a and if $\tilde{S} \in \gamma_a(S)$ we say that $\tilde{S}$ is a representative of $\gamma_a(S)$. Since $\{a\} \cap U = \{a\}$ and $\emptyset \cap U = \emptyset$ for all open sets U containing a, we write $\gamma_a(\{a\}) = \{a\}$ and $\gamma_a(\emptyset) = \emptyset$. If $a \notin \bar{S}$, $\gamma_a(S) = \emptyset$. The finite unions, intersections and complements of germs of sets at a are defined by the same operations on representatives. (It is easy to check that these are well-defined operations on germs, and independent of the chosen representatives.)

7.1 $\mathbb{F}$-ANALYTIC VARIETIES

DEFINITION 7.1.1 *Suppose that $U \subset \mathbb{F}^n$ is a non-empty open set and that G denotes a finite collection of functions $g : U \to \mathbb{F}$ which are $\mathbb{F}$-analytic on U. Let*

$$\mathrm{var}\,(U, G) = \{x \in U : g(x) = 0 \text{ for all } g \in G\}.$$

This is called the $\mathbb{F}$-analytic variety generated by G on U. If $U \subset \mathbb{C}^n$ and the elements of G are real-on-real, we say that $\mathrm{var}\,(U, G)$ is real-on-real provided that $U \cap \mathbb{R}^n \neq \emptyset$.

A point $x \in \mathrm{var}\,(U, G)$ is said to be m-regular if there is a neighbourhood O of x in $\mathbb{F}^n$ such that $O \cap \mathrm{var}\,(U, G)$ is an $\mathbb{F}$-analytic manifold of dimension m (see Definition 4.5.5). Note that

$$\mathrm{var}\,(U, G_1) \cap \mathrm{var}\,(U, G_2) = \mathrm{var}\,(U, G_1 \cup G_2)$$
$$\mathrm{var}\,(U, G_1) \cup \mathrm{var}\,(U, G_2) = \mathrm{var}\,(U, G_3)$$

where $G_3 = \{g_1 g_2 : g_i \in G_i, \, i = 1, 2\}$.

The germ at a of an $\mathbb{F}$-analytic variety is referred to as an $\mathbb{F}$-analytic germ and the germ of a real-on-real $\mathbb{C}$-analytic variety in $\mathbb{C}^n$ is called a real-on-real germ. The set of all $\mathbb{F}$-analytic germs at $a \in \mathbb{F}^n$ is denoted by $\mathcal{V}_a(\mathbb{F}^n)$. If $\alpha \in \mathcal{V}_a(\mathbb{F}^n)$, its dimension, $\dim_{\mathbb{F}} \alpha$, is the largest integer m such that every representative of α contains an m-regular point (the point a itself need not be m-regular.) If no such integer exists, we say that $\dim_{\mathbb{F}} \alpha = -1$.

REMARKS 7.1.2 For $a \in \mathbb{F}^n$, $\mathbb{F}^n$, $\{a\}$ and $\emptyset$ are elements of $V_a(\mathbb{F}^n)$ with $\dim_\mathbb{F} \emptyset = -1$, $\dim_\mathbb{F}\{a\} = 0$ and $\dim_\mathbb{F} \gamma_a(\mathbb{F}^n) = n$.

Theorem 6.2.5 says that $\gamma_0(\mathcal{A}) \in V_0(\mathbb{C}^m)$.

If $\alpha, \beta \in V_a(\mathbb{F}^n)$, then both $\alpha \cap \beta$ and $\alpha \cup \beta$ are in $V_a(\mathbb{F}^n)$, but in general $\alpha \setminus \beta \notin V_a(\mathbb{F}^n)$ □

LEMMA 7.1.3 *Suppose that $M \subset \mathbb{F}^n$ is an $\mathbb{F}$-analytic manifold (Definition 4.5.5) and $a \in M$. Then $\gamma_a(M) \in V_a(\mathbb{F}^n)$. If $a \in U \cap M$ and var (U, G) is an $\mathbb{F}$-analytic variety, there is an open neighbourhood W of a in M such that $W \setminus$ var (U, G) is either empty or dense in W.*

Proof. First we show that $\gamma_a(M)$ is in $V_a(\mathbb{F}^n)$. Without loss of generality suppose that $0 = a \in M$ and, in the notation of Definition 4.5.5, let $Z_1 =$ range $df[0]$, $\mathbb{F}^n = Z_1 \oplus Z_2$, and write

$$f(x) = f_1(x) + f_2(x) \in Z_1 \oplus Z_2, \quad x \in U_0 \subset \mathbb{F}^m,$$

where U_0 is a neighbourhood of $0 \in \mathbb{F}^m$. By hypothesis, $df[0]$ has rank m and $df_1[0] : \mathbb{F}^m \to Z_1$ is a bijection. By the analytic inverse function theorem 4.5.3, U_0 in Definition 4.5.5 can be chosen so that f_1, from $U_0 \subset \mathbb{F}^m$ onto a neighbourhood W_0 of $f_1(0) = 0 \in Z_1$, is a bijection with an analytic inverse. Now $\{f(x) : x \in U_0\}$ is a representative of $\gamma_0(M)$ in $\mathbb{F}^n$ and

$$\begin{aligned} \{f(x) : x \in U_0\} &= \{(f_1(x), f_2(x)) : x \in U_0\} \\ &= \{(y, f_2 \circ f_1^{-1}(y)) : y \in W_0\} \\ &= \{(y, z) \in W_0 \times Z_2 : z - f_2 \circ f_1^{-1}(y) = 0\}. \end{aligned}$$

Since $f_2 \circ f_1^{-1}$ is analytic this shows that $\gamma_0(M) \in V_0(\mathbb{F}^n)$.

Now suppose that var (U, G) is an $\mathbb{F}$-analytic variety in $\mathbb{F}^n$, let $0 \in M$ and let U_0 as above. Suppose that $B \subset U_0$ is a ball centred at $0 \in \mathbb{F}^m$ and let $W = f(B)$. Then $0 \in W$, which is a relatively open connected subset of M.

Suppose that $W \setminus$ var (U, G) is not dense in W. Then there is an open set $\hat{W} \subset W$ such that $\hat{W} \subset$ var (U, G). Let $\hat{B} = f^{-1}(\hat{W})$. Then $g \circ f \equiv 0$ on $\hat{B}$ for all $g \in G$. Hence $g \circ f \equiv 0$ on B for all $g \in G$, by Theorem 4.3.9. Hence $W \subset$ var (U, G), in other words, $W \setminus$ var (U, G) is empty. This proves the result. □

LEMMA 7.1.4 *Let var (U, G) be an $\mathbb{F}$-analytic variety in $\mathbb{F}^n$ and $M \subset U$ a connected $\mathbb{F}$-analytic manifold such that $M \cap$ var (U, G) has non-empty interior relative to M. Then $M \subset$ var (U, G).*

Proof. Let N^o denote the relative interior in M of $N = M \cap$ var (U, G). By definition, N^o is open in M, and non-empty by hypothesis. Suppose that x belongs to the boundary in M of N^o. By Lemma 7.1.3 there is an open neighbourhood W of x in M such that $W \setminus$ var (U, G) is either empty or dense in W. Now $W \cap N^o \neq \emptyset$ since x is on the boundary of N^o and, since $N^o \subset M \cap$ var (U, G) is open, $W \setminus$ var (U, G) is not dense in W. Hence it is empty, which implies that $x \in N^o$. Thus N^o is closed in M. By connectedness, $N^o = M$ and $M \subset$ var (U, G). □

DEFINITION 7.1.5 *A germ $\alpha \in \mathcal{V}_a(\mathbb{F}^n)$ is said to be irreducible if $\alpha = \alpha_1 \cup \alpha_2$ for germs $\alpha_1, \alpha_2 \in \mathcal{V}_a(\mathbb{F}^n)$ implies that $\alpha = \alpha_1$ or $\alpha = \alpha_2$.*

For example, $\emptyset$ and $\{a\}$ and are irreducible elements of $\mathcal{V}_a(\mathbb{F}^n)$.

LEMMA 7.1.6 *If M is an m-dimensional $\mathbb{F}$-analytic manifold and $a \in M$, then $\gamma_a(M) \in \mathcal{V}_a(\mathbb{F}^n)$ is irreducible.*

Proof. By Lemma 7.1.3, $\gamma_a(M) \in \mathcal{V}_a(\mathbb{F}^n)$. To see that it is irreducible suppose that E_1 and E_2 are $\mathbb{F}$-analytic varieties in $\mathbb{F}^n$ such that

$$\gamma_a(M) = \gamma_a(E_1 \cup E_2) \text{ and } \gamma_a(E_1) \neq \gamma_a(M) \neq \gamma_a(E_2).$$

It follows also from Lemma 7.1.3 that for $i = 1$, 2 there exists an open neighbourhood W of a in M such that $(M \setminus E_i) \cap W$ is either empty or dense in W. Note M and $E_1 \cup E_2$ coincide in a neighbourhood U of a in $\mathbb{F}^n$. Hence if $(M \setminus E_1) \cap W$ is empty, it follows that $E_2 \subset E_1$ in a neighbourhood of a and that $\gamma_a(M) = \gamma_a(E_1)$, which is assumed to be false. If $(M \setminus E_2) \cap W$ is empty we reach a similar contradiction.

Therefore $(M \setminus E_i) \cap W$ is dense, $i = 1$, 2 and so $(M \setminus (E_1 \cup E_2)) \cap W$ is dense in W. But this contradicts the fact that $\gamma_a(M) = \gamma_a(E_1 \cup E_2)$ and proves that $\gamma_a(M)$ is an irreducible germ. $\qquad\square$

We will see another important example of irreducible germs in Lemma 7.2.9. The most elementary non-trivial example of an analytic variety is one which is defined as the zeros in an open set $U \subset \mathbb{F}^n$, $n \geq 2$, of a single $\mathbb{F}$-analytic function $f : U \to \mathbb{F}$. Suppose, without loss of generality, that $0 \in U$, that $f(0) = 0$ and that $f \not\equiv 0$ on U. Then the Weierstrass preparation theorem 5.3.1 gives that there exists a choice of coordinates $(x_1, \cdots, x_n) \in \mathbb{F}^n$, $r > 0$ and an open set $V \subset \mathbb{F}^{n-1}$ containing 0 such that, with $U_0 = V \times B_r(\mathbb{F})$,

$$\text{var}(U_0, \{f\}) = \text{var}(U_0, \{h\}),$$
$$h(x_1, \cdots, x_n) = A(x_n; x_1, \cdots, x_{n-1}),$$

where $A(X; x_1, \cdots, x_{n-1})$ is a polynomial with coefficients that are analytic functions of $(x_1, \cdots, x_{n-1}) \in V$ of the form

$$A(X; x_1, \cdots, x_{n-1})$$
$$= X^q + \sum_{k=0}^{q-1} a_k(x_1, \cdots, x_{n-1}) X^k, \quad (x_1, \cdots, x_{n-1}) \in V,$$

with $a_k(0) = 0$, $0 \leq k \leq q - 1$. Since real polynomials need not have real roots, we need special structure to take this idea any further when $\mathbb{F} = \mathbb{R}$. However when $\mathbb{F} = \mathbb{C}$ polynomials do have roots in $\mathbb{C}$ and we have the following.

THEOREM 7.1.7 *When $\mathbb{F} = \mathbb{C}$ the polynomial A above can be chosen with the following additional properties:*

(a) *Its discriminant (see (6.3)) $D = D(a_0, \cdots, a_{q-1}, 1) \not\equiv 0$ on V.*

(b) *Every point of* var $(U_0, \{f\}) \setminus (\text{var}(V, \{D\}) \times \mathbb{C})$ *is an $(n-1)$-regular point of* var $(U_0, \{f\})$.

(c) $\dim_\mathbb{C} \alpha = n - 1$ *where α is the germ of* var $(U_0, \{f\})$.

(d) *If f is real-on-real, then A is real-on-real.*

Proof. (a) We simplify the polynomial obtained from the Weierstrass preparation theorem using Theorem 6.2.4. This may change the value of q and the coefficients a_k, but we retain the original notation for the simplified polynomial. Since, after simplification, $A(Z; z_1, \cdots, z_{n-1})$ has no multiple zeros for $(z_1, \cdots, z_{n-1})$ in an open dense connected subset of V (see Proposition 5.4.2), this proves (a).

(b) Since, for every $(z_1, \cdots, z_n) \in$ var $(U_0, \{f\}) \setminus$ var $((V, \{D\}))$, $Z = z_n$ is a simple zero of $A(Z; z_1, \cdots, z_{n-1})$, the analytic implicit function theorem 4.5.4 gives (b).

(c) Since $a_k(0) = 0$ and a_k is continuous on V, $0 \le k \le q - 1$, we know from Proposition 5.4.2 and Proposition 6.1.1 that there are points of var $(U_0, \{h\}) \setminus (\text{var}(V, \{D\}) \times \mathbb{C})$ arbitrarily close to zero. Therefore there are $(n-1)$-regular points of var $(U_0, \{f\})$ arbitrarily close to 0. Thus $\dim_\mathbb{C} \alpha = n - 1$.

(d) This is guaranteed by the Weierstrass preparation theorem 5.3.1 and Theorem 6.2.4. $\qquad\square$

REMARK 7.1.8   Of course, $\gamma_0(\text{var}(U, \{f\})) = \gamma_0(\text{var}(U_0, \{f\}))$. $\qquad\square$

Now we will develop the ideas involved in the proof of this result to obtain something much more general.

7.2 WEIERSTRASS ANALYTIC VARIETIES

Throughout $\mathbb{F} = \mathbb{C}$, $m \in \mathbb{N}_0$ and, for $m \in \mathbb{N}$, $V \subset \mathbb{C}^m$ is given by

$$V = \{(z_1, \cdots, z_m) \in \mathbb{C}^m : |z_k| < \delta, \ 1 \le k \le m\}.$$

In the case $m = 0$, $V = \{0\}$.

DEFINITION 7.2.1   *When $m \in \mathbb{N}$, a Weierstrass polynomial on V is a polynomial $A(Z; z_1, \cdots, z_m)$, $(z_1, \cdots, z_m) \in V$, of the form*

$$Z^p + \sum_{k=0}^{p-1} a_k(z_1, \cdots, z_m)Z^k, \quad p \in \mathbb{N}, \tag{7.1}$$

where

$$a_0(0) = \cdots = a_{p-1}(0) = 0 \text{ and } D(a_0, \cdots, a_{p-1}, 1) \not\equiv 0 \text{ on } V.$$

(By Proposition 5.4.2, $D(a_0, \cdots, a_{p-1}, 1) \neq 0$ on a connected, open, dense subset of V.) When $m = 0$ Weierstrass polynomials are of the form Z^p, $p \in \mathbb{N}$.

REMARK 7.2.2   Suppose that the coefficients a_k in a polynomial of the form
(7.1) vanish at $0 \in V$ and the discriminant is identically zero on V. Then its simplification, Theorem 6.2.4, is a Weierstrass polynomial on V. (All its coefficients,
apart from the principal coefficient, are zero at $0 \in V$.)

If A, B are Weierstrass polynomials on V and C is any (non-constant in Z)
polynomial on V with $AC = B$, then C is a Weierstrass polynomial.   □

Let $n > m \in \mathbb{N}_0$ and consider a family $\{A_{m+1}, \cdots, A_n\}$ of Weierstrass polynomials on V. For each $k \in \{m+1, \cdots, n\}$ let

$$h_k(z_1, \cdots, z_n) = A_k(z_k; z_1, \cdots, z_m).$$

Let H denote the family of $n - m$ functions h_k defined in this way. (Each $h_k \in H$
is a polynomial in z_k with coefficients that are analytic functions on $V \subset \mathbb{C}^m$.
Thus h_k is independent of all z_j for $j \in \{m+1, \cdots, n\} \setminus \{k\}$.)

DEFINITION 7.2.3   *A Weierstrass analytic variety is a set in $\mathbb{C}^n$ of the form*
var $\left(V \times \mathbb{C}^{n-m}, H\right)$, $0 \leq m < n$. *For our purposes, a Weierstrass analytic variety
is identified with the set H of Weierstrass analytic polynomials which define it and,
if $m \in \mathbb{N}$, its discriminant $D(H) : V \to \mathbb{C}$ is the product of the discriminants of
the polynomials A_k used in the definition.*

For $m \in \mathbb{N}$, the branches of a Weierstrass analytic variety var $\left(V \times \mathbb{C}^{n-m}, H\right)$
are the components of

$$\text{var} \left(V \times \mathbb{C}^{n-m}, H\right) \setminus \left(\text{var}\left(V, D(H)\right) \times \mathbb{C}^{n-m}\right).$$

REMARKS 7.2.4   All points on a branch of a Weierstrass analytic variety var $\left(V \times \mathbb{C}^{n-m}, H\right)$, for $n \in \mathbb{N}$ and $0 < m < n$, are m-regular because, by the analytic
implicit function theorem 4.5.4, in a neighbourhood of such a point each of the co-
ordinates $z_{m+1}, \cdots, z_n$ depends locally analytically on $(z_1, \cdots, z_m) \in V$. Thus
each branch is a connected $\mathbb{C}$-analytic manifold of dimension m and by Proposition
5.4.2 it projects onto the connected set $V \setminus$ var $\left(V, \{D(H)\}\right)$. But it is not possible
in general to define any one of the coordinates $z_{m+1}, \cdots, z_n$ as an analytic function on $V \setminus$ var $(V, \{D(H)\})$ when the latter set is multiply connected. In the proof
of Corollary 7.4.4 we will see that $E = $ var $\left(V \times \mathbb{C}^{n-m}, H\right)$ contains no manifold
of dimension strictly greater that m and hence $\dim_{\mathbb{C}} \gamma_0(E) = m$.

When $m = 0$, the only Weierstrass analytic variety in $\mathbb{C}^n$ is $\{0\}$ since all Weierstrass polynomials are of the form Z^p, $p \in \mathbb{N}$.   □

The following are a few elementary examples.

EXAMPLES 7.2.5   In Definition 7.2.1 with $m = 1$, $n = 2$ and $V = \mathbb{C}$, the polynomial $A(Z; z_1) = Z^2 - z_1$, $z_1 \in \mathbb{C}$, defines a Weierstrass analytic variety which
has exactly one branch, $B = \{(z_1, z_2) : z_2^2 = z_1, z_1 \neq 0\}$. This illustrates both
that a branch is connected, but not in general simply connected, and that "above"
each point of V there is usually more than one point of B. Note also that A is
real-on-real and $B \cap \mathbb{R}^2$ is a real parabola with the origin removed.

Let $E = \text{var}(\mathbb{C} \times \mathbb{C}, \{h\})$, $h(z_1, z_2) = A(z_2; z_1)$ and $A(Z; z) = Z^2 + z^2$. Note that $D(\{h\})$ is zero only at $0 \in \mathbb{C}$, that E has two branches, $B_\pm = \{(z, \pm iz) : z \in \mathbb{C}\}$ and that neither of them is closed under complex conjugation even though E is real-on-real. The next result addresses this issue. $\qquad\square$

LEMMA 7.2.6 *Suppose that B is a branch of a real-on-real $\mathbb{C}$-analytic variety* $\text{var}(V \times \mathbb{C}^{n-m}, H)$ *and that $B \cap \mathbb{R}^n \neq \emptyset$. Then*

$$B^* := \{(\bar{z}_1, \cdots, \bar{z}_n) : (z_1, \cdots, z_n) \in B\} = B.$$

Proof. Since functions in H have Taylor expansions at 0 with real coefficients and B is a maximal connected set (in $(V \setminus \text{var}(V, \{D(H)\})) \times \mathbb{C}^{n-m}$) of solutions of the equations $h = 0$, $h \in H$, the set B^* is a maximal in the same sense. By hypothesis, $B \cap B^* \neq \emptyset$. Hence $B^* = B$ and the result is proved. $\qquad\square$

The next result shows that the closure of a branch is an analytic variety in a neighbourhood of 0, even though the branch, itself a manifold, need not be. For a branch B,

$$\overline{B} \text{ denotes } \overline{B} \cap (V \times \mathbb{C}^{n-m}), \text{ the relative closure of } B \text{ in } V \times \mathbb{C}^{n-m}.$$

By Proposition 5.4.2, a Weierstrass analytic variety is the union of the closures (in this sense) of its branches.

THEOREM 7.2.7 *Suppose that B is a branch of a Weierstrass analytic variety* $E = \text{var}(V \times \mathbb{C}^{n-m}, H)$ *with discriminant $D = D(H)$. Then*

$$\overline{B} = \text{var}(V \times \mathbb{C}^{n-m}, G)$$

for some finite collection of analytic functions $g : V \times \mathbb{C}^{n-m} \to \mathbb{C}$.

Suppose in addition that H is real-on-real and $B \cap \mathbb{R}^n \neq \emptyset$. Then G is real-on-real, $\overline{B}$ is a real-on-real $\mathbb{C}$-analytic variety and $\overline{B} \cap \mathbb{R}^n$ is an $\mathbb{R}$-analytic variety with $\dim_{\mathbb{R}} \gamma_0(\mathbb{R}^n \cap \overline{B}) = m$.

Proof. Let $\xi = (z_1, \cdots, z_m) \in V \setminus \text{var}(V, \{D\})$. Then there are K, say, points of B above ξ. In other words, there are K elements $\zeta_j(\xi) \in \mathbb{C}^{n-m}$, such that $(\xi, \zeta_j(\xi)) \in B$, $1 \leq j \leq K$. By Remarks 7.2.4 the dependence of $\zeta_j(\xi)$ on ξ is $\mathbb{C}$-analytic locally and therefore, by connectedness, K is independent of $\xi \in V \setminus \text{var}(V, \{D\})$. Note also that $(\xi, \zeta) \in B$ if and only if, for all $\varrho = (\varrho_{m+1}, \cdots, \varrho_n) \in \mathbb{C}^{n-m}$

$$\prod_{j=1}^{K} \langle \varrho, \zeta - \zeta_j(\xi) \rangle = 0.$$

Since this product, as a function of $(\xi, \zeta) \in (V \setminus \text{var}(V, \{D\})) \times \mathbb{C}^{n-m}$, is independent of permutations of $j \in \{1, 2, \cdots, K\}$, it is a continuous single-valued function of $(\xi, \zeta) \in \left(V \setminus \text{var}\,(V, \{D\})\right) \times \mathbb{C}^{n-m}$ and therefore is $\mathbb{C}$-analytic there. Therefore

$$\prod_{j=1}^{K} \langle \varrho, \zeta - \zeta_j(\xi) \rangle = \sum_{\substack{\sigma \in \mathbb{N}_0^{n-m} \\ |\sigma|=K}} \varrho^\sigma \tilde{g}_\sigma(\xi, \zeta), \tag{7.2}$$

where the functions $\tilde{g}_\sigma$ are analytic on $\left(V \setminus \text{var}\,(V, \{D\})\right) \times \mathbb{C}^{n-m}$. Moreover, for $\xi \in V \setminus \text{var}\,(V, \{D\})$,

$$(z_1, \cdots, z_n) = (\xi, \zeta) \in B \text{ if and only if } \tilde{g}_\sigma(z_1, \cdots, z_n) = 0$$

for all $\sigma \in \mathbb{N}_0^{n-m}$ with $|\sigma| = K$.

Finally observe that, for all compact sets $W \subset V$,

$$\sup\{|\zeta_j(\xi)| : \xi \in W \setminus \text{var}\,(V, \{D\}), \, 1 \le j \le K\} < \infty.$$

Therefore $\tilde{g}_\sigma$ is bounded on $(W \times \mathbb{C}^{n-m}) \setminus \text{var}\,(V \times \mathbb{C}^{n-m}, \{D\})$ and, by the Riemann extension theorem 5.4.3, can be extended as an analytic function g_σ on all of $V \times \mathbb{C}^{n-m}$. To complete the proof of the first part let $G = \{g_\sigma : \sigma \in \mathbb{N}_0^{n-m}, |\sigma| = K\}$ and recall that $V \setminus \text{var}\,(V, \{D\})$ is open, dense and connected in V.

Now suppose that H is real-on-real and $B \cap \mathbb{R}^n \neq \emptyset$. From the implicit function theorem 3.5.4 with $\mathbb{F} = \mathbb{R}$, $B \cap \mathbb{R}^n$ is an $\mathbb{R}$-analytic manifold of dimension m. By Lemma 7.2.6, for each j, $\zeta_j(\xi) = \overline{\zeta_k(\xi)}$ for some k. Therefore the left side of (7.2) is real when ϱ, ξ and ζ are real vectors. Therefore $\tilde{g}_\sigma(\xi, \zeta)$ is real when ξ and ζ are real vectors. This shows that G is real-on-real. Therefore E is a real-on-real $\mathbb{C}$-analytic variety and $\dim_{\mathbb{R}} B \cap \mathbb{R}^n = m$. $\qquad\square$

REMARK 7.2.8 Suppose that $m = n - 1$ in the preceding theorem. From (7.2) it follows that G has only one element, g say, where g is a polynomial in z_n with coefficients analytic on V, its principal coefficient is 1, all the others vanish at $0 \in V$ and its discriminant is not identically zero. Therefore in Theorem 7.2.7 $\overline{B} = \text{var}\,(V \times \mathbb{C}^{n-1}, G)$ is a Weierstrass analytic variety on V.

The following example of a Weierstrass analytic variety in $\mathbb{C}^3$ with $m = n - 2$ shows that this observation may be false when $m \neq n - 1$. Let $V = \mathbb{C}$ and let $E = \text{var}\,(V \times \mathbb{C}^2, \{h, k\})$ where

$$h(x, y) = y^2 - x^3, \quad k(x, z) = z^2 - x^3, \quad (x, y, z) \in \mathbb{C}^3.$$

Note that E has two branches $B_\pm$,

$$\overline{B}_\pm = \text{var}\,(V \times \mathbb{C}^2, \{h, k, l^\pm\}), \text{ where } l^\pm = yz \pm x^3, \quad (x, y, z) \in \mathbb{C}^3. \tag{7.3}$$

We now show that neither is a Weierstrass analytic variety on V. Suppose that this is false and that $\overline{B}_-$ is a Weierstrass analytic variety defined by

$$y^p + \sum_{k=0}^{p-1} A_k(x) y^k = 0 \text{ and } z^q + \sum_{l=0}^{q-1} B_l(x) z^l = 0, \tag{7.4}$$

where the discriminant of the polynomials is non-zero almost everywhere. Therefore, for almost all x in a neighbourhood of 0 in $\mathbb{C}$, there are exactly pq solutions of (7.4). However, for the same x there are two points (x, y, z) on $\overline{B}_-$. Hence $pq = 2$. Suppose $p = 1$ and $q = 2$. Then the system

$$y = A_0(x) \text{ and } z^2 = B_1(x)z + B_0(x)$$

is equivalent to (7.3) with a minus sign. But this is false since (7.3) does not determine y as a function of x. A similar contradiction is reached if $q = 1$ and $p = 2$, and for $\overline{B}_+$. $\qquad\square$

We have seen in Definition 7.2.3, Remark 7.2.4 and Lemma 7.1.6 that $\gamma_a(B)$ is irreducible when $a \in B$ and B is a branch of a Weierstrass analytic variety. More is true.

LEMMA 7.2.9 *Let B be a branch of a Weierstrass analytic variety* var $(V \times \mathbb{C}^{n-m}, H)$. *Then $\gamma_0(\overline{B}) \in \mathcal{V}_0(\mathbb{C}^n)$ is irreducible.*

Proof. The case $m = 0$ is trivial since $B = \{0\}$. Suppose $m \in \mathbb{N}$. Suppose that $\gamma_0(\overline{B}) = \alpha_1 \cup \alpha_2, \alpha_1, \alpha_2 \in \mathcal{V}_0(\mathbb{C}^n)$. Let E_1, E_2 be analytic varieties, defined in a neighbourhood of $0 \in \mathbb{C}^n$, which represents α_1 and α_2. Then there exists an open set O in $\mathbb{C}^n$ with $0 \in O$ such that $\overline{B} \cap O = O \cap (E_1 \cup E_2)$.

Now for any $z \in B \cap O$, $\gamma_z(B) \in \mathcal{V}(\mathbb{C}^n)$ is irreducible, by Lemma 7.1.6. Therefore for every point $z \in B \cap O$ there is an open set O_z in $\mathbb{C}^n$ with $O_z \cap B \subset E_1$ or $O_z \cap B \subset E_2$. Since B is a connected analytic manifold it follows from Lemma 7.1.4 that $B \subset E_1$ or $B \subset E_2$. Since E_i is closed in $V \times \mathbb{C}^{n-m}$, $\overline{B} \subset E_1$ or $\overline{B} \subset E_2$. Hence $\gamma_0(\overline{B}) = \gamma_0(E_1) = \alpha_1$ or $\gamma_0(\overline{B}) = \gamma_0(E_2) = \alpha_2$, and $\gamma_0(\overline{B})$ is irreducible. This completes the proof. $\qquad\square$

In Remark 7.2.8, the Weierstrass analytic variety E is defined in terms of Weierstrass polynomials h, k neither of which is the product of two Weierstrass polynomials on V, yet E is not irreducible. The following shows that this does not happen when $m = n - 1$.

LEMMA 7.2.10 *Let E denote a Weierstrass analytic variety*

$$\text{var } (V \times \mathbb{C}, \{h\}), \quad h(z_1, \cdots, z_n) = A(z_n; z_1, \cdots, z_{n-1}).$$

Then $\gamma_0(E)$ is irreducible if and only if A is not the product of two Weierstrass polynomials on V.

Proof. Suppose that $\gamma_0(E)$ is not irreducible. By Lemma 7.2.9, E has at least two branches and by Remark 7.2.8 the closure of each of these branches is a Weierstrass analytic variety. Suppose one such branch $\tilde{B}$ has closure defined by a Weierstrass polynomial $\tilde{A}$ on V, where $\tilde{A}$ and A are distinct polynomials on V. Then the greatest common divisor of A and $\tilde{A}$ is $\tilde{A}$ and $A = \tilde{A}E$ for some non-trivial Weierstrass polynomial E on V (see the last sentence of Remark 7.2.2). Therefore A is the product of two Weierstrass polynomials.

When A is the product $A_1 A_2$ of two Weierstrass polynomials, it is easy to see that $\gamma_0(E) = \gamma_0(E_1) \cup \gamma_0(E_2)$, where E_i are the varieties defined using A_j. Since A is a Weierstrass polynomial, $A_1 \neq A_2$ and therefore $\gamma_0(E)$ is not irreducible. This completes the proof. □

7.3 ANALYTIC GERMS AND SUBSPACES

Suppose that α is a $\mathbb{C}$-analytic germ at a with $\gamma_a(\mathbb{C}^n) \neq \alpha$. Suppose that var (U, G) is a representative of α. Then there is at least one $\mathbb{C}$-analytic function $g \in G$, such that $g \not\equiv 0$ on U. Hence there is a complex line segment L through a in U such that $g \not\equiv 0$ on L. Since g restricted to L is a complex-analytic function of one complex variable, its zeros are isolated. This shows that there exists a one-dimensional complex linear space Y such that $\gamma_a(a + Y) \cap \alpha = \{a\}$.

For the case of real-on-real varieties var (U, G), suppose that $0 \in U$ and that $g \in G$ is real-on-real. Then the coefficients in the Taylor expansion of g at 0 are all real and not all zero. Hence there exists a real linear space $\hat{Y} = \{tb : t \in \mathbb{R}\}$, for some $b \in \mathbb{R}^n$, such that $\gamma_0(\hat{Y}) \cap \alpha = \{0\}$. Moreover, g is not identically zero on the complex linear space $T = \{zb : z \in \mathbb{C}\}$ and hence $\gamma_0(T) \cap \alpha = \{0\}$. We will say that a complex linear subspace T of $\mathbb{C}^n$ is a complexified subspace if it has a real basis (see Remark 5.1.2). Equivalently T is complexified if it is closed under complex conjugation of the coordinates of its vectors with respect to the standard real basis. Clearly $T = \{zb : z \in \mathbb{C}\}$, $b \in \mathbb{R}^n$, is a complexified space of one complex dimension. Any complexified subspace Z_1 of $\mathbb{C}^n$ has a (in fact many) complementary complexified subspace Z_2 of $\mathbb{C}^n$ such that $\mathbb{C}^n = Z_1 \oplus Z_2$. This ensures that the choice of basis in the last part of the next lemma is possible.

LEMMA 7.3.1 *Suppose that $\alpha \in V_0(\mathbb{C}^n)$, $n \geq 2$, and Y is a linear subspace of $\mathbb{C}^n$ such that $\alpha \cap \gamma_0(Y) = \{0\}$. Choose a basis of $\mathbb{C}^n$ such that*

$$Y = \{(0, \cdots, 0, z_{m+1}, \cdots, z_n) : (z_{m+1}, \cdots, z_n) \in \mathbb{C}^{n-m}\} \qquad (7.5)$$

and let P denote the projection onto $\mathbb{C}^m$ given by first m coordinates, so that

$$P(E) = \{(z_1, \cdots, z_m) : (z_1, \cdots z_n) \in E\}, \qquad (7.6)$$

where E is a representative of α. Then $\gamma_0(P(E)) \in V(\mathbb{C}^m)$.

If Y is a complexified subspace, α is real-on-real and we choose a real basis of $\mathbb{C}^n$ such that (7.5) holds. Then $\gamma_0(P(E))$ is real-on-real.

Proof. The cases $m = 0$ ($\alpha = \gamma_0(\mathbb{C}^n)$, $Y = \{0\}$) and $m = n$ ($\alpha = \{0\}$, $Y = \mathbb{C}^n$) are trivial and we suppose throughout that $0 < m < n$. In the coordinates (7.5), let $\alpha = \gamma_0(\text{var}(W \times B_\delta(\mathbb{C})), \{g_1, \cdots, g_\nu\})$, $\nu \in \mathbb{N}$, for a small positive δ where

$$W = \{(z_1, \cdots, z_{n-1}) \in \mathbb{C}^{n-1} : |z_j| < \delta, \ 1 \leq j \leq n-1\}.$$

Since

$$\{(0, \cdots, 0, z) \in \mathbb{C}^n : z \in \mathbb{C}\} \subset Y \text{ and } \gamma_0(Y) \cap \alpha = \{0\},$$

$g_k(0, \cdots, 0, z_n) \not\equiv 0$ for some k when $z_n \in B_\delta(\mathbb{C})$. Relabelling $\{g_1, \cdots, g_\nu\}$ if necessary, suppose that $k = 1$. By the Weierstrass preparation theorem 5.3.1, there is no loss of generality in supposing that g_1 is given by a polynomial A_1 of the form (5.1).

If A_1 has degree p, say, then, by the Weierstrass division theorem 5.2.1, we may suppose without loss of generality that each of the other g_k, $k \geq 2$, in the definition of α, has the form

$$g_k(z_1, \cdots, z_n) = A_k(z_n; z_1, \cdots, z_{n-1})$$

where $A_k(Z; z_1, \cdots, z_{n-1})$ is a polynomial on W of degree at most $p - 1$. Thus var $(W \times B_\delta(\mathbb{C}), \{g_1, \cdots, g_\nu\})$ is a representative of α and the family $\{g_1, \cdots, g_\nu\}$ of analytic functions satisfies the hypotheses of the projection lemma, Theorem 6.2.5.

Therefore the projection of var $(W \times B_\delta(\mathbb{C}), \{g_1, \cdots, g_\nu\})$ onto $\mathbb{C}^{n-1} \times \{0\}$ is an analytic variety in $\mathbb{C}^{n-1}$. Let $\beta \in \gamma_0(\mathbb{C}^{n-1})$ denote its germ and let

$$\hat{Y} = \{(0, \cdots, 0, z_{m+1}, \cdots, z_{n-1}) : (z_{m+1}, \cdots, z_{n-1}) \in \mathbb{C}^{n-m-1}\}.$$

Since $\alpha \cap \gamma_0(Y) = \{0\}$ in $\mathbb{C}^n$, $\gamma_0(\hat{Y}) \cap \beta = \{0\}$ in $\mathbb{C}^{n-1}$. We can now repeat the argument $n - m$ times to prove the first part of the lemma.

In the case when Y is a complexified subspace, choose a real basis for $\mathbb{C}^n$ such that (7.5) holds. Then the projection lemma at each step gives a real-on-real variety. This completes the proof. $\square$

LEMMA 7.3.2 *Let Y be a linear subspace which is maximal with respect to $\alpha \in \mathcal{V}_0(\mathbb{C}^n)$ in the sense that for any linear subspace $\hat{Y}$ of $\mathbb{C}^n$*

$$\gamma_0(Y) \cap \alpha = \{0\} \text{ and } \hat{Y} \neq Y \subset \hat{Y} \text{ implies that } \gamma_0(\hat{Y}) \cap \alpha \neq \{0\}. \tag{7.7}$$

Let $n \geq 2$, $m = n - \dim Y \in \{1, \cdots, n-1\}$, and choose coordinates (7.5). Let E be any representative of α. Then $\gamma_0(P(E)) = \gamma_0(\mathbb{C}^m)$, where $P(E)$ is defined in (7.6).

Proof. We have seen in the previous lemma that $P(E)$ is an analytic variety in $\mathbb{C}^m$. Suppose that at 0 its germ $\beta \neq \gamma_0(\mathbb{C}^m)$. Then there exists a non-trivial linear space $L \subset \mathbb{C}^m$ such that $\beta \cap \gamma_0(L) = \{0\}$. Let $\hat{Y} = L \times Y$. Then $\gamma_0(\hat{Y}) \cap \alpha = \{0\}$ which violates the maximality of Y. This proves the lemma. $\square$

This is false for real analytic germs as the real germ of $\{(x, y) \in \mathbb{R}^2 : y - x^2 = 0\}$ illustrates.

LEMMA 7.3.3 *Suppose that $\alpha \in \mathcal{V}_0(\mathbb{C}^n)$ is real-on-real, $n \geq 2$, and T is a complexified subspace of $\mathbb{C}^n$ such that $\alpha \cap \gamma_0(T) = \{0\}$. Suppose that T is a maximal complexified subspace in the sense that if $\tilde{T}$ is a complexified space then*

$$\gamma_0(T) \cap \alpha = \{0\} \text{ and } T \subset \tilde{T} \neq T \text{ implies that } \gamma_0(\tilde{T}) \cap \alpha \neq \{0\}. \tag{7.8}$$

With respect to a real basis such that

$$T = \{(0, \cdots, 0, z_{m+1}, \cdots, z_n) : (z_{m+1}, \cdots, z_n) \in \mathbb{C}^{n-m}\}, \qquad (7.9)$$

$\gamma_0(P(E)) = \gamma_0(\mathbb{C}^m)$.

Proof. We have seen in Lemma 7.3.1 that in this case $P(E)$ is a real-on-real analytic variety in $\mathbb{C}^m$. Suppose that its germ at 0, $\beta \neq \gamma_0(\mathbb{C}^m)$. Then, by the remarks at the beginning of the section, there exists a non-trivial complexified subspace $L \subset \mathbb{C}^m$ such that $\beta \cap \gamma_0(L) = \{0\}$. Let $\tilde{T} = L \times T$. Then $\gamma_0(\tilde{T}) \cap \alpha = \{0\}$ which violates the maximality of T. This proves the lemma. □

EXAMPLE 7.3.4 The following is an illustration of (7.7) and (7.8). Let $g, h : \mathbb{C}^4 \to \mathbb{C}$ be defined by

$$g(w, x, y, z) = y^2 + z^2, \quad h(w, x, y, z) = x, \quad (w, x, y, z) \in \mathbb{C}^4.$$

Then E is real-on-real and $E \cap \mathbb{R}^4 = \text{span}_{\mathbb{R}}\{(1, 0, 0, 0)\}$ where

$$E = \text{var}(\mathbb{C}^4, \{g, h\}) = \{(w, 0, y, z) \in \mathbb{C}^4 : y^2 + z^2 = 0\},$$

and $\alpha \cap \gamma_0(\{0\} \times \mathbb{R}^3) = \{0\}$, where $\alpha = \gamma_0(E)$. However the three-dimensional complexified space $T = \{0\} \times \mathbb{C}^3$ is not maximal in the sense of (7.8); it is too big. It is easy to see that each of the two-dimensional complexified spaces $\{0\} \times \mathbb{C} \times \{0\} \times \mathbb{C}$ and $\{0\} \times \mathbb{C} \times \mathbb{C} \times \{0\}$ are maximal in the sense of (7.7) and (7.8). □

7.4 GERMS OF $\mathbb{C}$-ANALYTIC VARIETIES

We are now in a position to show that if α is a $\mathbb{C}$-analytic germ at 0 there exists a Weierstrass analytic variety E, a subset C and a branch B of E such that $\alpha = \gamma_0(C)$ and $\gamma_0(B) \subset \alpha$.

Suppose that $\alpha \in \mathcal{V}_0(\mathbb{C}^n)$. If $n = 1$ then $\alpha \in \{\emptyset, \{0\}, \gamma_0(\mathbb{C})\}$. If $\{0\} \subset \alpha$ but $\alpha \notin \{\{0\}, \gamma_0(\mathbb{C}^n)\}$ then $n \geq 2$. Moreover, since $\alpha \neq \gamma_0(\mathbb{C}^n)$, §7.3 ensures the existence of a non-trivial linear subspace Y of $\mathbb{C}^n$ such that

$$\gamma_0(Y) \cap \alpha = \{0\} \qquad (7.10)$$

and, since $\alpha \neq \{0\}$, we infer that $Y \neq \mathbb{C}^n$. Let Y be such a linear subspace, $n \geq 2$,

$$m = n - \dim Y \in \{1, \cdots, n-1\} \qquad (7.11)$$

and choose coordinates such that

$$Y = \{(0, \cdots, 0, z_{m+1}, \cdots, z_n) : (z_{m+1}, \cdots, z_n) \in \mathbb{C}^{n-m}\}. \qquad (7.12)$$

THEOREM 7.4.1 *Let $\alpha \in \mathcal{V}_0(\mathbb{C}^n) \setminus \{\{0\}, \gamma_0(\mathbb{C}^n)\}$, $n \geq 2$, and choose coordinates such that (7.10), (7.11) and (7.12) hold. Then there exists a Weierstrass analytic variety* var $(V \times \mathbb{C}^{n-m}, H)$, $1 \leq m < n$, *such that*

$$\alpha \subset \gamma_0(\text{var}(V \times \mathbb{C}^{n-m}, H)).$$

Proof. As in the proof of Lemma 7.3.1, the Weierstrass preparation 5.3.1 and division 5.2.1 theorems, and Proposition 6.2.4 can be used to reduce the problem to the case when $\alpha = \gamma_0(\mathrm{var}\,(W \times \mathbb{C}, G))$, where $G = \{g_1, \cdots, g_\nu\}$, $\nu \in \mathbb{N}$, and

$$g_1(z_1, \cdots, z_n)$$
$$= z_n^p + a_{p-1}(z_1, \cdots, z_{n-1})z_n^{p-1} + \cdots + a_0(z_1, \cdots, z_{n-1}),$$

$$a_j(0) = 0, \quad 0 \le j \le p-1, \quad D(g_1) \not\equiv 0 \text{ on } W,$$

and each g_k, $k \ge 2$, is a polynomial in the same variable z_n, with coefficients that are $\mathbb{C}$-analytic functions of the other variables; g_1 is the only polynomial in G with highest degree p.

First let $\nu = 1$ and $G = \{g_1\}$ where $2 \le n \in \mathbb{N}$ is arbitrary. Since g_1 is a Weierstrass polynomial in z_n, Proposition 6.1.1 gives that

$$\alpha = \gamma_0\big(\{(z_1, \cdots, z_{n-1}) \in \mathbb{C}^{n-1}$$
$$: (z_1, \cdots, z_n) \in \mathrm{var}\,(W \times \mathbb{C}, \{g_1\})\}\big) = \gamma_0(\mathbb{C}^{n-1}).$$

By (7.10) and (7.12), $m = n - 1$ and the theorem holds with $H = \{g_1\}$ and $V = W$. (In fact we get $\alpha = \gamma_0(\mathrm{var}\,(V \times \mathbb{C}, H))$ in this case.)

For the general case when $G = \{g_1, \cdots, g_\nu\}$ we argue by induction on $n \ge 2$. The inductive hypothesis is that for all $n \ge 3$ and all $\hat{\alpha} \in \mathcal{V}_0(\mathbb{C}^{\hat{n}})$, $2 \le \hat{n} < n$, the conclusion of the theorem holds with m, n, Y replaced with $\hat{m}$, $\hat{n}$, $\hat{Y}$ satisfying

$$\gamma_0(\hat{Y}) \cap \hat{\alpha} = \{0\}, \quad \hat{n} \ge 2, \hat{m} = \hat{n} - \dim \hat{Y} \in \{1, \cdots, \hat{n} - 1\}$$
$$\hat{Y} = \{(0, \cdots, 0, z_{\hat{m}+1}, \cdots, z_{\hat{n}}) : (z_{\hat{m}+1}, \cdots, z_{\hat{n}}) \in \mathbb{C}^{\hat{n}-\hat{m}}\}.$$

This hypothesis has been verified when $n = 3$ because then $\hat{n} = 2$, $\hat{m} = 1$. We no longer need to consider the case $\alpha = \gamma_0(\mathrm{var}\,(W \times \mathbb{C}, \{g_1\}))$. So suppose $n \ge 3$ and, for some $\nu \ge 2$,

$$\alpha = \gamma_0\big(\mathrm{var}\,(W \times \mathbb{C}, \{g_1, \cdots, g_\nu\})\big),$$

where the set $G = \{g_1, \cdots, g_\nu\}$ satisfies the hypotheses of the projection lemma, Theorem 6.2.5. Let $\hat{\alpha} = \gamma_0(\mathcal{A})$, where $\mathcal{A} \subset W \subset \mathbb{C}^{n-1}$ denotes the set given by the projection lemma, and note that $\hat{\alpha} \in \mathcal{V}_0(\mathbb{C}^{n-1})$ by Remark 7.1.2.

Suppose that $\hat{\alpha} = \{0\}$. Then in a sufficiently small neighbourhood of the origin in $\mathbb{C}^n$,

$$(z_1, \cdots, z_n) \in \mathrm{var}\,(W \times \mathbb{C}, G) \text{ only if } z_n^p = g_1(0, 0, \cdots, z_n) = 0.$$

It now follows that $\alpha = \{0\}$ which contradicts the hypothesis of the theorem. Hence $\hat{\alpha} \ne \{0\}$. Now suppose that $\hat{\alpha} = \gamma_0(\mathbb{C}^{n-1})$. It follows from (7.10) and (7.12) that $n - 1 = m$. Let $H = \{g_1\}$, $V = W$. Since $\alpha \subset \gamma_0(\mathrm{var}\,(V \times \mathbb{C}, H))$, the theorem holds in this case.

Finally we come to the case $\hat{\alpha} \notin \{\{0\}, \gamma_0(\mathbb{C}^{n-1})\}, n \geq 3, m < n - 1$. Let

$$\hat{Y} = \{(z_1, \cdots, z_{n-1}) : z_1 = \cdots = z_m = 0\},$$

where m is defined in (7.10) and (7.12). It follows from the definition of $\mathcal{A}$ and (7.12) that $\gamma_0(\hat{Y}) \cap \hat{\alpha} = \{0\}$. With $\hat{n} = n - 1$, $\hat{m} = m$, the inductive hypothesis gives that the theorem holds in $\mathbb{C}^{n-1}$, $n \geq 3$. Thus, in the same coordinates, there exists (a possibly smaller) $\delta > 0$, a set

$$V = \{(z_1, \cdots, z_m) \in \mathbb{C}^m : |z_1|, \cdots, |z_m| < \delta\},$$

and a collection $\hat{H} = \{\hat{A}_{m+1}, \cdots, \hat{A}_{n-1}\}$ of Weierstrass polynomials on V with discriminant $\hat{D} = D(H)$ not identically zero and

$$\gamma_0(\mathcal{A}) \subset \gamma_0(V \times \mathbb{C}^{n-m-1}, \hat{H}).$$

Let

$$\Upsilon(z_1, \cdots, z_m) = \{(\hat{z}_{m+1}, \cdots, \hat{z}_{n-1}) \subset \mathbb{C}^{n-m-1} : \\ \hat{A}_j(\hat{z}_j; z_1, \cdots, z_m) = 0, \ m + 1 \leq j \leq n - 1\}.$$

Since $\hat{D} \not\equiv 0$, the dependence of the $\hat{z}_{m+j}$ on $(z_1, \cdots, z_m) \in V \setminus \mathrm{var}\,(V, \hat{D})$, an open, dense, connected subset of V, is locally $\mathbb{C}$-analytic, by the analytic implicit function theorem 4.5.4. Now define a polynomial on $V \setminus \mathrm{var}\,(V, \hat{D})$ by

$$\hat{A}_n(Z; z_1, \cdots, z_m)$$

$$= \prod_{\substack{(\hat{z}_{m+1}, \cdots, \hat{z}_{n-1}) \\ \in \Upsilon(z_1, \cdots, z_m)}} A_1(Z; z_1, \cdots, z_m, \hat{z}_{m+1}, \cdots, \hat{z}_{n-1}).$$

By choosing a smaller value of δ in the definition of V if necessary we see that the coefficients of $\hat{A}_n$ are bounded and hence, by the Riemann extension theorem 5.4.3, can be extended as $\mathbb{C}$-analytic functions to all of V. Note that in $\hat{A}_n$ the coefficient of the highest power of Z is 1 and that all the other coefficients vanish at $0 \in V$. After simplification (Remark 7.2.2) $\hat{A}_n$ becomes a Weierstrass polynomial on V.

Let $H = \hat{H} \cup \{\hat{A}_n\}$, a collection of $n - m$ Weierstrass polynomials on V. Let $D(z_1, \cdots, z_m)$ denote the product of their discriminants, which is non-zero on an open dense connected subset of V.

Now $\mathrm{var}\,(V \times \mathbb{C}^{n-m}, H)$ is a Weierstrass analytic variety. Suppose $(z_1, \cdots, z_n)$ belongs to a representative of α in a sufficiently small neighbourhood of 0. Then

$$(z_1, \cdots, z_{n-1}) \in \mathcal{A}, \quad (z_1, \cdots, z_m) \in V,$$
$$(z_{m+1}, \cdots, z_{n-1}) \in \Upsilon(z_1, \cdots, z_m),$$
$$g_1(z_1, \cdots, z_n) = 0.$$

Thus

$$\alpha \subset \gamma_0\big(\mathrm{var}\,(V \times \mathbb{C}^{n-m}, H)\big).$$

This completes the proof.   □

THEOREM 7.4.2 *Let* $\alpha = \gamma_0(E)$ *in the preceding theorem and with* P *as in Lemma 7.3.1 suppose* $\gamma_0(P(E)) = \gamma_0(\mathbb{C}^m)$. *Then* $\gamma_0(B) \subset \alpha$ *for some branch* B *of the Weierstrass analytic variety* var $(V \times \mathbb{C}^{n-m}, H)$ *in Theorem 7.4.1.*

Proof. Without loss of generality suppose that $E = $ var $(V \times \mathbb{C}^{m-n}, G)$, where G is a finite collection of analytic functions and $V \times \mathbb{C}^{n-m}$ is as in Theorem 7.4.1. With H as in the conclusion of this theorem, let $O \subset V \setminus $ var $(V, (D(H)))$ be a non-empty open ball on which the discriminant $D(H)$ is nowhere zero. For any $\xi \in O$, we may write

$$\left(\{\xi\} \times \mathbb{C}^{n-m}\right) \cap \text{var } (V \times \mathbb{C}^{n-m}, H) = \{(\xi, \zeta_j(\xi)) : 1 \leq j \leq p\},$$

where p is the product of the degrees of the Weierstrass analytic polynomials in H, and the ζ_j are analytic functions from O into $\mathbb{C}^{n-m}$, as in Remark 7.2.4. According to Lemma 7.1.3, every open set $O_j = \{\xi \in O : (\xi, \zeta_j(\xi)) \notin E\}$ is either empty or dense in O. However, by hypothesis, $\cap_{j=1}^p O_j$ is empty. Therefore at least one of these sets, O_{j_0}, is empty. In other words the analytic manifold

$$M_{j_0} = \{(\xi, \zeta_{j_0}(\xi)) : \xi \in O_{j_0}\}$$

is a subset of E. By Lemma 7.1.4, $\left(B \cap (V \times \mathbb{C}^{n-m})\right) \subset E$, where B is the branch which contains M_{j_0} and the proof is complete.   □

COROLLARY 7.4.3 *From the maximality hypotheses of Lemmas 7.3.2 and 7.3.3, the conclusion on Theorem 7.4.2 holds.*

Proof. This follows by combining the lemmas and the theorem cited.   □

COROLLARY 7.4.4 (Dimension of α**)** *(a) Suppose* $n \geq 2$, *the hypotheses of Lemma 7.3.2 hold, and consequently, in the notation of Theorem 7.4.2,*

$$\gamma_0(B) \subset \alpha \subset \gamma_0\left(\text{var } (V \times \mathbb{C}^{n-m}, H)\right). \tag{7.13}$$

Also, var $\left(V \times \mathbb{C}^{n-m}, H\right)$ *contains no manifold of dimension larger than* $m = \dim_{\mathbb{C}} \alpha$, *and*

$$n - m = \max\{\dim Y : \gamma_0(Y) \cap \alpha = \{0\}, Y \subset \mathbb{C}^n \text{ a linear space}\}. \tag{7.14}$$

The right hand side of (7.14) is called the codimension of the analytic germ α.
Moreover var $(V \times \mathbb{C}^{n-m}, H \cup \{D(H)\})$ *contains no manifold of dimension equal to or greater that* m.
(b) If the hypotheses of Lemma 7.3.3 hold and m *is defined there instead, then* $m = \dim_{\mathbb{C}} \alpha$ *and the conclusion of part (a) is valid. In this case the dimension of* α *is equal to* m *whether defined in terms of maximal complex subspaces as in part (a), or of maximal complexified spaces as in part (b).*

Proof. It is clear from (7.13) that $\dim_{\mathbb{C}} B = m \leq \dim_{\mathbb{C}} \alpha$. If $D(H)$ is nowhere zero, or zero only at $0 \in V$, it is easy to see that $E = \mathrm{var}\,(V \times \mathbb{C}^{n-m}, H)$ contains no manifold of dimension larger than m and hence $\dim_{\mathbb{C}} \alpha = m$. Suppose that $\{0\} \subset \gamma_0(\mathrm{var}\,(V, \{D(H)\})) \notin \{\{0\}, \gamma_0(\mathbb{C}^m)\}$ and let N be a manifold of dimension strictly greater than m which is a subset of E. If $x \in B \cap N$ for some branch B then a neighbourhood of x in N is a subset of a neighbourhood of x in B, which is impossible since B is an m-dimensional manifold. Therefore $B \cap N = \emptyset$ for all branches B of E. In other words $N \subset \mathrm{var}\,(V \times \mathbb{C}^{n-m}, H \cup D(H))$.

Let $P_j(\xi)$ denote the jth coordinate of $\xi \in \mathbb{C}^n$ or $\mathbb{C}^m$. Then P_j is an analytic function and §7.3 gives the existence of coordinates on V such that

$$\gamma_0(V, \{D(H), P_1, \cdots, P_{m-1}\}) = \{0\},$$

so that

$$\gamma_0\left(\mathrm{var}\,(V \times \mathbb{C}^{n-m}, H \cup \{D(H)\})\right) \cap \gamma_0(Y) = \{0\}$$

where

$$Y = \{(z_1, \cdots, z_n) \in \mathbb{C}^n : z_1 = \cdots = z_{m-1} = 0\}.$$

Therefore Theorem 7.4.1 yields the existence of a Weierstrass analytic variety $\mathrm{var}\,(\tilde{V} \times \mathbb{C}^{n-\tilde{m}}, \tilde{H}))$, $\tilde{m} < m$, such that

$$N \subset \mathrm{var}\,(V \times \mathbb{C}^{n-m}, H \cup \{D(H)\}) \subset \mathrm{var}\,(\tilde{V} \times \mathbb{C}^{n-\tilde{m}}, \tilde{H}).$$

Repeated finitely often we find that this holds with $\tilde{m} = 1$, which is impossible. Hence $\mathrm{var}\,(V \times \mathbb{C}^{n-m}, H)$ contains no analytic manifold of dimension larger than m, $\dim_{\mathbb{C}} \alpha = m$ and $\mathrm{var}\,(V \times \mathbb{C}^{n-m}, H \cup \{D(H)\})$ contains no manifold of dimension m or more. That (7.14) holds follows from Corollary 7.4.3.

The proof of (b) is the same once Lemma 7.3.3 is taken into account.   □

Now we improve slightly on the observation in Lemma 7.2.9 that $\gamma_0(\overline{B})$ is irreducible when B is a branch of a Weierstrass analytic variety.

LEMMA 7.4.5   *Let B be a branch of a Weierstrass analytic variety* $\mathrm{var}\,(V \times \mathbb{C}^{n-m}, H)$ *and suppose that $\alpha \in V_0(\mathbb{C}^n)$ is such that $\gamma_0(\overline{B}) \neq \alpha \subset \gamma_0(\overline{B})$. Then $\dim_{\mathbb{C}} \alpha < m$.*

Proof. Suppose that $\gamma_0(\overline{B}) \neq \alpha \subset \gamma_0(\overline{B})$, and let $E = \mathrm{var}\,(U, G) \subset \overline{B} \cap (V \times \mathbb{C}^{n-m})$, where U is an open set with $B \cup \{0\} \subset U$ and such that $\alpha = \gamma_0(E)$. Let $D(H)$ denote the discriminant of H on V and suppose that $\dim_{\mathbb{C}} \alpha \geq m$. We will infer that $\gamma_0(\overline{B}) = \alpha$, a contradiction which will prove the lemma.

Define an analytic manifold $M \subset E$ as consisting of all $(\dim_{\mathbb{C}} \alpha)$-regular points of E. If $M \subset \mathrm{var}\,(V \times \mathbb{C}^{n-m}, H \cup \{D(H)\})$, then $\dim_{\mathbb{C}} \alpha < \dim_{\mathbb{C}} B = m$, by Lemma 7.4.4. Since this is false, by assumption, $M \cap B \neq \emptyset$ and $\dim_{\mathbb{C}} \alpha = \dim_{\mathbb{C}} M = m$. Therefore there exists a point $z \in M \cap B$ which has a neighbourhood O_z in B which is a subset of M. From Lemma 7.1.4 and the fact that B is connected it follows that $B \subset E$. This contradiction proves the result.   □

COROLLARY 7.4.6 *Suppose that $m = 1$ in Theorem 7.4.1. Then*

$$\alpha = \gamma_0\left(\{0\} \cup_{\gamma_0(B) \subset \alpha} B\right),$$

where the union is over all branches B of var $(V \times \mathbb{C}^{n-1}, H)$ *with germ contained in α.*

Proof. Since $m = 1$, $V \subset \mathbb{C}$ and there is no loss of generality in assuming that the discriminant $D(H)$ is non-zero on $V \setminus \{0\}$ and that $\overline{B} = B \cup \{0\}$ for all branches B. Now suppose that B is a branch of var $(V \times \mathbb{C}^{n-1}, H)$ such that $\alpha \cap \gamma_0(B) \neq \{0\}$ and that $\gamma_0(B) \not\subset \alpha$. Then $\gamma_0(\overline{B}) \not\subset \alpha$ and $\alpha \cap \gamma_0(\overline{B})$ is an analytic germ in $\mathbb{C}^n$ with

$$\gamma_0(\overline{B}) \neq \alpha \cap \gamma_0(\overline{B}) \subset \gamma_0(\overline{B}).$$

Therefore, by Lemma 7.4.5, $\dim_{\mathbb{C}} \left(\alpha \cap \gamma_0(\overline{B})\right) = 0$. Hence $\alpha \cap \gamma_0(\overline{B}) = \{0\}$. Since this is false we conclude that $\gamma_0(B) \subset \alpha$ for every branch B of var $(V \times \mathbb{C}, H)$ with $\gamma_0(B) \cap \alpha \not\subset \{0\}$. This proves the corollary. $\square$

We are now in a position to say something, but not everything, about the structure of $\mathbb{C}$-analytic varieties. The following extension of the preceding corollary is more than sufficient for our purposes.

THEOREM 7.4.7 **(A Structure Theorem)** *Let $n \geq 2$ and $\alpha \in \mathcal{V}_0(\mathbb{C}^n) \setminus \{0\}$ be such that $\{0\} \subset \alpha \neq \gamma_0(\mathbb{C}^n)$. Then there exist sets $B_1, \cdots, B_N$, such that*

(a) $\alpha = \gamma_0\left(B_1 \cup \cdots \cup B_N \cup \{0\}\right).$

(b) *Each B_j, $1 \leq j \leq N$, after a linear change of coordinates (depending on j), is a branch of a Weierstrass analytic variety (depending, including its dimension, on j).*

(c) $\dim_{\mathbb{C}} \alpha = \max_{1 \leq j \leq N}\{\dim_{\mathbb{C}} B_j\}.$

(d) *If $L \subset \mathbb{C}^n$, $\gamma_0(L) \neq \emptyset$, is a connected $\mathbb{C}$-analytic manifold of dimension $l \in \{1, \cdots, n\}$ the points of which are l-regular points of a representative of α, then there exists $j \in \{1, \cdots, N\}$ such that $\gamma_0(L) \subset \gamma_0(\overline{B_j})$ and $\dim_{\mathbb{C}} B_j = l$.*

(e) *If α is real-on-real, then it can be arranged that each branch B_j with $B_j \cap \mathbb{R}^n \neq \emptyset$ is real-on-real.*

(f) $\alpha \cap \gamma_0(\mathbb{R}^n) = \gamma_0\left(\tilde{B}_1 \cup \cdots \cup \tilde{B}_K \cup \{0\}\right)$ *where the $\tilde{B}_j$ denotes those branches which intersect $\mathbb{R}^n$ non-trivially.*

(g) $\dim_{\mathbb{R}}(\alpha \cap \mathbb{R}^n) = \max_{1 \leq j \leq K} \dim_{\mathbb{R}}(\tilde{B}_j \cap \mathbb{R}^n).$

Proof. We use induction on the dimension of α. Suppose that var $(V \times \mathbb{C}^{n-m}, H)$ and the coordinate system are given by Theorem 7.4.2. Consider first all the branches of var $(V \times \mathbb{C}^{n-m}, H)$ such that $\gamma_0(B) \subset \alpha$. If $m = 1$, Corollary 7.4.6 shows that these are all the branches that we need for the result to hold.

Suppose $m \geq 2$ and make the inductive hypothesis that the results (a)-(d) of the theorem hold for all smaller values of m.

(a)-(c) According to Theorem 7.4.1,

$$\alpha \subset \gamma_0\big(\text{var}\,(V \times \mathbb{C}^{n-m}, H)\big) = \cup\overline{B},$$

where the union is over all branches B of var $(V \times \mathbb{C}^{n-m}, H)$. In addition to the branches B with $\gamma_0(B) \subset \alpha$, consider branches $\tilde{B}$ of var $(V \times \mathbb{C}^{n-m}, H)$ such that

$$\emptyset \neq \gamma_0(\tilde{B}) \cap \alpha \neq \gamma_0(\tilde{B}).$$

Since, by Lemma 7.4.5, the germ $\alpha \cap \gamma_0(\overline{\tilde{B}})$ has dimension strictly smaller than m, we can apply the inductive hypothesis to each of these branches to complete the proof of (a)-(c).

(d) Note that $\dim_{\mathbb{C}} L \leq m = \dim_{\mathbb{C}} \alpha$ and that we may suppose $L \subset V \times \mathbb{C}^{n-m}$. If $\gamma_0(L) \subset \gamma_0\big(\text{var}\,(V \times \mathbb{C}^{n-m}, H \cup \{D(H)\})\big)$, we apply the inductive hypothesis to obtain the required result.

If $\gamma_0(L) \not\subset \gamma_0\big(\text{var}\,(V \times \mathbb{C}^{n-m}, H \cup \{D(H)\})\big)$, then $L \cap B \neq \emptyset$ for at least one branch B of var $(V \times \mathbb{C}^{n-m}, H)$. For all $z \in L \cap B$, L and B coincide locally, in a neighbourhood of z and $\dim_{\mathbb{C}} L = \dim_{\mathbb{C}} B$. By Theorem 7.2.7 $\overline{B} \cap (V \times \mathbb{C}^{n-m})$ is an analytic variety. That $\gamma_0(L) \subset \gamma_0(\overline{B})$ now follows from Lemma 7.1.4.

(e) It is clear that if α is real-on-real, and maximal complexified spaces are used, as in Lemma 7.3.3, to choose coordinates, then the branches B_j which emerge are real-on-real. Parts (f) and (g) follow from the second part of Theorem 7.2.7. □

COROLLARY 7.4.8 *Let* $\alpha \in \mathcal{V}_0(\mathbb{C}^n) \setminus \{\emptyset, \{0\}, \gamma_0(\mathbb{C}^m)\}$ *be irreducible. Then (possibly after a linear change of coordinates)* $\alpha = \gamma_0(\overline{B})$ *where B is a branch of some Weierstrass analytic variety. If α is real-on-real and $\alpha \cap \gamma_0(\mathbb{R}^n) \neq \{0\}$, then B is a branch of a real-on-real variety.*

Proof. In the notation of Theorem 7.4.7, $\alpha = \gamma_0(\overline{B_1}) \cup \cdots \cup \gamma_0(\overline{B_N})$, and since α is irreducible the result follows. □

The following example shows that when α is an irreducible $\mathbb{C}$-analytic variety it does not necessarily follow that $\alpha \cap \gamma_0(\mathbb{R}^n)$ is an irreducible real analytic variety.

EXAMPLE 7.4.9 Let $V = \mathbb{C}^2$ and $E = \text{var}\,(V \times \mathbb{C}, \{h\})$ where

$$h(x, y, z) = z^2 + x^2 y^2 (x^2 + y^2), \quad (x, y, z) \in V \times \mathbb{C}.$$

Clearly $E \subset \mathbb{C}^3$ is a Weierstrass analytic variety defined by the polynomial

$$A(Z; x, y) = Z^2 + x^2 y^2 (x^2 + y^2), \quad (x, y, z) \in V \times \mathbb{C}.$$

If A is a product of two Weierstrass polynomials then each has order one, and it is easily checked that this is impossible. Hence, by Lemma 7.2.10, $\gamma_0(E)$ is an irreducible $\mathbb{C}$-analytic variety. However $E \cap \mathbb{R}^3$ coincides with

$$\{(x, y, z) \in \mathbb{R}^3 : x = 0,\ z = 0\} \cup \{(x, y, z) \in \mathbb{R}^3 : y = 0,\ z = 0\}.$$

Therefore $\gamma_0(E \cap \mathbb{R}^n)$ is not a real irreducible variety.   □

7.5 ONE-DIMENSIONAL BRANCHES

The following results are aspects of the theory of Puiseux series sufficient for our later needs.

THEOREM 7.5.1   *Suppose that $m = 1$, $2 \le n \in \mathbb{N}$ and that B is a branch of the Weierstrass analytic variety $E = \mathrm{var}\,(V \times \mathbb{C}^{n-1}, H)$ where V is chosen so that $D(H)$ is non-zero on $V \setminus \{0\}$. Then there exist $K \in \mathbb{N}$, $\delta > 0$ and a $\mathbb{C}$-analytic function*

$$\psi : \{z \in \mathbb{C} : |z|^K < \delta\} \to \mathbb{C}^{n-1}$$

such that the mapping $z \mapsto (z^K, \psi(z))$ is injective, $\psi(0) = 0$ and

$$\{0\} \cup B = \overline{B} \cap (V \times \mathbb{C}^{n-1}) = \{(z^K, \psi(z)) : |z|^K < \delta\}.$$

REMARK 7.5.2   A function ψ satisfying the conclusion of the theorem is not unique. Indeed, if ψ satisfies the theorem and $\zeta \neq 1$ is a Kth root of unity then $\tilde{\psi}(z) := \psi(\zeta z)$ defines another function which also satisfies the conclusion of the theorem.   □

Proof. Let $H = \{h_2, \cdots, h_n\}$ where $h_k(z_1, \cdots, z_n) = A_k(z_k; z_1)$, and each A_k is a Weierstrass polynomial of degree p_k, say, $2 \le k \le n$. If the discriminant $D(H)$ is not zero at $z_1 = 0$, then, for all $k \in \{2, \cdots, n\}$, $A_k(Z; z_1) = Z - a_k(z_1)$ where a_k, is an analytic function on $V \subset \mathbb{C}$. In this case the theorem holds with $K = 1$ and

$$\psi(z_1) = (a_2(z_1), a_3(z_1), \cdots, a_n(z_1)), \quad z_1 \in V.$$

Now suppose that $D(H)$ is zero at 0. Note that for $z_1 \in V \setminus \{0\}$ each of the polynomials $A_k(Z; z_1)$ has only simple roots. Let $\hat{V}$ denote the half-plane in $\mathbb{C}$ defined by

$$\hat{V} = \{z \in \mathbb{C} : z = \rho + i\theta,\ -\infty < \rho < \log \delta,\ \theta \in \mathbb{R}\},$$

where δ is given in the definition of V, and let

$$\hat{h}_k(z, z_k) = A_k(z_k; e^z), \quad z \in \hat{V},\ z_k \in \mathbb{C}.$$

Let

$$\hat{H} = \{\hat{h}_2, \cdots, \hat{h}_n\} \text{ and } \hat{E} = \text{var}\,(\hat{V} \times \mathbb{C}^{n-1}, \hat{H}).$$

It is clear that B is a branch of E if and only if $\hat{B}$ is a branch of $\hat{E}$, where

$$B = \{(e^z, \xi) : (z, \xi) \in \hat{B}\}, \quad \xi \in \mathbb{C}^{n-1}.$$

Since $D(H)$ is nowhere zero on $V \setminus \{0\}$, $D(\hat{H})$ is nowhere zero on $\hat{V}$ and every point of $\hat{E}$ is 1-regular (Definition 7.1.1). We can therefore write

$$(\{z\} \times \mathbb{C}^{n-1}) \cap \hat{E} = \{(z, \xi_q(z)) : 1 \le q \le p\},$$

where $p = \prod_{k=2}^{n} p_k$. By the analytic implicit function Theorem 4.5.4, each ξ_q is defined locally on $\hat{V}$ as a $\mathbb{C}$-analytic function with values in $\mathbb{C}^{n-1}$ and, since $\hat{V}$ is simply connected, they define analytic functions on $\hat{V}$. Thus $\hat{E}$ is the union of the disjoint graphs of the functions $\xi_q : \hat{V} \to \mathbb{C}^{n-1}$, $1 \le q \le p$.

Recall that, for $z \in \hat{V}$, each component of $\xi_q(z) \in \mathbb{C}^{n-1}$ is a simple root of a polynomial $A_k(Z; e^z)$, $2 \le k \le n$. Therefore the set-valued map

$$z \mapsto \{(e^z, \xi_q(z)) : 1 \le q \le p\}$$

is $2\pi i$-periodic on $\hat{V}$. Moreover if, for some $\hat{z} \in \hat{V}$ and some $m \in \mathbb{Z}$,

$$\xi_{q_1}(\hat{z}) = \xi_{q_2}(\hat{z} + 2\pi m i), \quad q_1, q_2 \in \{1, \cdots p\},$$

then

$$\xi_{q_1}(z) = \xi_{q_2}(z + 2\pi m i) \text{ for all } z \in \hat{V},$$

by the analytic implicit function theorem 4.5.4 and analytic continuation. Therefore, for each $q \in \{1, \cdots, p\}$, the mapping

$$z \mapsto (e^z, \xi_q(z)) \in E, \ z \in \hat{V}, \tag{7.15}$$

is periodic with period $2\pi K_q i$ and is injective on the set $V_q = \{z = \rho + i\theta \in \hat{V} : 0 < \theta \le 2\pi K_q\}$, for some $K_q \in \{1, \cdots, p\}$. It is easy to see that its image on V_q is both open and closed in E and hence is a branch of E.

For a given branch B, choose q such that the image of (7.15) on V_q coincides with B. We have seen that an injective parameterization of B is given by

$$B = \{(e^z, \xi_q(z)) : z \in V_q\}.$$

Since $z \mapsto \xi_q(K_q z)$ has period (not necessarily minimal) $2\pi i$, we can define an analytic function $\tilde{\psi} : \{z : 0 < |z| < \delta^{1/K_q}\} \to \mathbb{C}$ by

$$\tilde{\psi}(z_1) = \xi_q(K_q \log z_1)$$

where it does not matter which branch of log is chosen. Thus

$$\xi_q(K_q z) = \tilde{\psi}(e^z), \ K_q z \in \hat{V}.$$

This gives a new injective parameterization of B, namely

$$B = \{(z_1^{K_q}, \tilde{\psi}(z_1)) : 0 < |z| < \delta^{1/K_q}\},$$

where ψ is analytic and $\lim_{z_1 \to 0} \tilde{\psi}(z_1) = 0$. The Riemann extension theorem 5.4.3 means that $\tilde{\psi}$ has an analytic extension ψ defined on the ball $\{z_1 \in \mathbb{C} : |z_1| < \delta^{1/K_q}\}$ with $\psi(0) = 0$. Let $K = K_q$ to complete the proof. $\qquad\square$

COROLLARY 7.5.3   *In Theorem 7.5.1 suppose $\gamma_0(B \cap \mathbb{R}^n) \notin \{\emptyset, \{0\}\}$. Then there exists $k \in \mathbb{N}_0$ with $0 \le k \le 2K - 1$ such that*

$$\mathbb{R}^n \cap \overline{B} = \{((-1)^k r^K, \psi(r \exp(k\pi i/K))) : -\delta^{1/K} < r < \delta^{1/K}\}, \qquad (7.16)$$

and this parameterization is injective.

Proof. Since $\gamma_0(B \cap \mathbb{R}^n) \notin \{\{0\}, \emptyset\}$ there exists, by Theorem 7.5.1, a sequence $\{z_j\} \subset \mathbb{C}$ with $z_j \to 0$ such that $z_j^K \in \mathbb{R}$ and $\psi(z_j) \in \mathbb{R}^{n-1}$ for all $j \in \mathbb{N}$. Therefore, without loss of generality we may assume, for some $k \in \{0, 1, \cdots, 2K - 1\}$, that $z_j = |z_j| \exp(k\pi i/K)$ and

$$\psi(|z_j| \exp(k\pi i/K)) \in \mathbb{R}^{n-1} \text{ for all } j \in \mathbb{N}.$$

Since ψ is a $\mathbb{C}$-analytic function of one complex variable we can infer that

$$\psi(r \exp(k\pi i/K)) \in \mathbb{R}^{n-1} \text{ for all } r \in \mathbb{R} \text{ with } -\delta^{1/K} < r < \delta^{1/K}.$$

If there exists $l \in \{0, 1, \cdots, 2K - 1\}$ different from k and a sequence $\rho_j > 0$ such that $\psi(\rho_j \exp(l\pi i/K))$ is real and $\rho_j \to 0$ as $j \to \infty$, then, by the preceding argument, we may assume that $\psi(r \exp(l\pi i/K))$ is real for all r with $-\delta^{1/K} < r < \delta^{1/K}$. We will now show that $l - k \in K\mathbb{Z}$.

Suppose that this is false. For $p \in \mathbb{N}$,

$$\frac{d^p \psi}{dr^p} (r \exp(k\pi i/K)) \bigg|_{r=0}$$

$$= \exp(p\pi i(k - l)/K) \frac{d^p \psi}{dr^p} (r \exp(l\pi i/K)) \bigg|_{r=0},$$

and the derivatives are real. Therefore, for all p with $p(l - k) \notin K\mathbb{Z}$, it follows that

$$\frac{d^p \psi}{dz^p}(0) = 0.$$

Let $p_0 \in \mathbb{N}$ be the generator of the ideal $\{p \in \mathbb{Z} : p(l - k) \in K\mathbb{Z}\}$. Then the power series expansion of $\psi(z)$ at $z = 0$ involves only powers of z^{p_0} and it follows that

$$\psi(z_1) = \psi(z_2)$$

for all $z_1, z_2 \in \mathbb{C}$ such that $z_1^{p_0} = z_2^{p_0}$. Since p_0 divides K, $z_1^K = z_2^K$ for all such z_1, z_2. Therefore if $z_1^{p_0} = z_2^{p_0}$,

$$\left(z_1^K, \psi(z_1)\right) = \left(z_2^K, \psi(z_2)\right).$$

Now the injectivity in the theorem above gives that $p_0 = 1$ and $k - l \in K\mathbb{Z}$. This completes the proof. □

LEMMA 7.5.4 *Suppose the Weierstrass polynomials which define the Weierstrass analytic variety* var $(V \times \mathbb{C}^{n-1}, H)$ *in Theorem 7.5.1 are real-on-real and that the discriminant $D(H)$ is non-zero on $V \setminus \{0\} \subset \mathbb{C}$. Then $B \cap \mathbb{R}^n \notin \{\emptyset, \{0\}\}$ implies that $\gamma_0(B \cap \mathbb{R}^{n-1}) \notin \{\emptyset, \{0\}\}$.*

Proof. Suppose that $(\hat{x}_1, \hat{x}_2, \cdots, \hat{x}_n) \in (\mathbb{R}^n \cap B) \setminus \{0\}$. Then each $\hat{x}_k$, $k \geq 2$, is a simple root, when $z_1 = \hat{x}_1 \in \mathbb{R}$, of a polynomial whose coefficients ($\mathbb{C}$-analytic functions of z_1) are real when $z_1 = x_1 \in \mathbb{R}$, and are zero when $z_1 = 0$. From the real implicit function theorem 3.5.4 it follows that each of the polynomials in H has a real root $x_k \in \mathbb{R}$, $2 \leq k \leq m$, which is a real-valued analytic function of x_1 when $\hat{x}_1 x_1 > 0$ and x_1 is sufficiently small. Moreover, $x_k \to 0$ as $x_1 \to 0$ by Lemma 6.1.1. It follows from Theorem 7.5.1 that there exists $k \in \{0, 1, \cdots, 2K - 1\}$ such that

$$\psi(r \exp(k\pi i/K)) \in \mathbb{R}^{n-1} \quad \text{for} \quad -\delta^{1/K} < r < \delta^{1/K}.$$

□

EXAMPLE 7.5.5 Consider the collection H of three Weierstrass analytic polynomials

$$Z^2 - z_1, \quad Z^3 - z_1^2, \quad Z^4 - z_1^3,$$

and let V be the disc of radius 2 with centre 0 in $\mathbb{C}$. Then the corresponding Weierstrass analytic variety,

$$\text{var } (V \times \mathbb{C}^3, H)$$
$$= \{(z_1, z_2, z_3, z_4) : |z_1| < 2, z_2^2 - z_1 = z_3^3 - z_1^2 = z_4^4 - z_1^3 = 0\},$$

has two branches. To see this note that the branch which contains $(1, 1, 1, 1)$ also contains the closed Jordan curve

$$\Gamma_1 = \{(e^{it}, e^{(it/2)}, e^{(2it/3)}, e^{(3it/4)}) : t \in [0, 24\pi]\}.$$

Γ_1 projects onto the unit circle in V and contains 12 points above $1 \in \mathbb{C}$. Similarly the branch B_2 containing $(1, -1, 1, 1)$ contains the closed Jordan curve

$$\Gamma_2 = \{(e^{it}, e^{i(\pi + t/2)}, e^{(2it/3)}, e^{(3it/4)}) : t \in [0, 24\pi]\}.$$

Γ_2 also has 12 points above $1 \in \mathbb{C}$ and projects onto the unit circle in V. Since the equations

$$t = s \mod 2\pi, \qquad\qquad \pi + t/2 = s/2 \mod 2\pi,$$
$$2t/3 = 2s/3 \mod 2\pi, \qquad\qquad 3t/4 = 3s/4 \mod 2\pi,$$

imply that

$$2k = 4l - 2 = 3m = 8n/3 \text{ for some } k, l, m, n \in \mathbb{Z},$$

they have no solutions. Therefore $\Gamma_1 \cap \Gamma_2 = \emptyset$. Since there are at most $2 \times 3 \times 4 = 24$ points of var $(V \times \mathbb{C}^3, H)$ above $1 \in V$ there are at most two branches of this variety. Thus the variety has exactly two branches and $K = 12$ in Theorem 7.5.1. Moreover candidates for the function ψ (remember it is not unique) corresponding to these branches, are

$$\psi_1(z) = (z^6, z^8, z^9) \text{ and } \psi_2 = (z^6, z^8, -iz^9).$$

Moreover

$$B_1 \cap \mathbb{R}^n = \{(r^{12}, r^6, r^8, r^9) : r \in (-2^{1/12}, 2^{1/12})\}$$
$$B_2 \cap \mathbb{R}^n = \{(r^{12}, -r^6, r^8, r^9) : r \in (-2^{1/12}, 2^{1/12})\},$$

which correspond to $k = 0$ and $k = 6$, respectively, in Corollary 7.5.3.

Note that Lemma 7.5.4 is false if the coefficients of elements of H are complex when their argument is real. For example, when $n = 2$ let $h(Z, z_1) = Z^2 - (1 + i)z_1 - iz_1^2$. Then var $(V, \{h\}) \cap \mathbb{R}^2 = \{(0,0), (1,1), (1,-1)\}$. $\qquad\square$

7.6 NOTES ON SOURCES

This material is to be found in the books Chirka [18], Federer [29], Mumford [45] and Narasimhan [46]. For the theory of analytic varieties far beyond our needs, see Lojasiewicz [43].

PART 3
Bifurcation Theory

Chapter Eight

Local Bifurcation Theory

In this chapter we consider the existence of solutions to nonlinear equations of the form $F(\lambda, x) = 0$ where $F : \mathbb{F} \times X \to Y$, $F(\lambda, 0) = 0 \in Y$ for all $\lambda \in \mathbb{F}$ and X and Y are Banach spaces. A solution is a pair (λ, x) and the parameter is not prescribed *a priori*. For the moment we treat the cases of differentiable F and analytic F simultaneously.

Local bifurcation theory addresses the question: for which $\lambda_0 \in \mathbb{F}$ is there a sequence $\{(\lambda_n, x_n)\} \subset \mathbb{F} \times (X \setminus \{0\})$ of solutions converging to $(\lambda_0, 0)$ in $\mathbb{F} \times X$? Such $(\lambda_0, 0)$ are called bifurcation points on the line of trivial solutions $\{(\lambda, 0) : \lambda \in \mathbb{F}\}$. For obvious reasons, $\lambda_0 \in \mathbb{F}$ is sometimes referred to as the bifurcation point.

8.1 A NECESSARY CONDITION

Suppose that F is continuously differentiable from a neighbourhood of $(\lambda_0, 0)$ in $\mathbb{F} \times X$ into Y. If $\partial_x F[(\lambda_0, 0)]$ is a homeomorphism, the implicit function Theorem 3.5.4 says that in a neighbourhood of $(\lambda_0, 0)$ in $\mathbb{F} \times X$ all the solutions to $F(\lambda, x) = 0$ lie on a unique curve $\{(\lambda, x) : x = \phi(\lambda), \ \lambda \in (\lambda_0 - \epsilon, \lambda_0 + \epsilon)\}$, for some $\epsilon > 0$. Since the line of trivial solutions passes through $(\lambda_0, 0)$, we conclude that $\phi(\lambda) = 0$ for all $\lambda \in (\lambda_0 - \epsilon, \lambda_0 + \epsilon)$.

Hence, when F is continuously differentiable at $(\lambda_0, 0)$, a necessary condition for $(\lambda_0, 0)$ to be a bifurcation point is that $\partial_x F[(\lambda_0, 0)] : X \to Y$ should not be a homeomorphism. Note that, because X and Y are Banach spaces and since, by definition, $\partial_x F[(\lambda_0, 0)] : X \to Y$ is bounded, this is equivalent to the weaker statement that $\partial_x F[(\lambda_0, 0)] : X \to Y$ should not be a bijection (see Corollary 2.4.3). This latter condition is often much easier to verify in particular situations.

EXAMPLE 8.1.1 Suppose $X = Y$ and that the C^1-function F has the form $F(\lambda, x) = x - G(\lambda, x)$, where $G(\lambda, \cdot) : X \to Y$ is a compact nonlinear operator with $G(\lambda, 0) = 0$ for all $\lambda \in \mathbb{F}$. Now

$$\partial_x F[(\lambda_0, 0)] = I - \partial_x G[(\lambda_0, 0)],$$

where $\partial_x G[(\lambda_0, 0)]$ is a compact linear operator on X by Lemma 3.1.12. Therefore, by the Fredholm alternative 2.7.5, $\partial_x F[(\lambda_0, 0)]$ is a homeomorphism if and only if $\ker \partial_x F[(\lambda_0, 0)] = \{0\}$ and a necessary condition for $(\lambda_0, 0)$ to be a bifurcation point is that the linear equation $\partial_x F[(\lambda_0, 0)]x = 0$ should have at least one non-trivial solution.

More generally, when $\partial_x F[(\lambda_0, 0)]$ is Fredholm with index zero (see §2.7),

$$\{0\} \neq \ker \partial_x F[(\lambda_0, 0)] : X \to Y$$

is a necessary condition for $(\lambda_0, 0)$ to be a bifurcation point. $\qquad \square$

The following example shows that this condition is not sufficient.

EXAMPLE 8.1.2 Let $X = Y = \mathbb{C}$, regarded as Banach spaces over $\mathbb{R}$, and let $F(\lambda, z) = z - \lambda z - i|z|^2 z$. Then $\partial_z F[(\lambda_0, 0)]z = (1 - \lambda_0)z$ and hence $\lambda_0 = 1$ satisfies the condition which is necessary for $(\lambda_0, 0)$ to be a bifurcation point. However, $F(\lambda, z) = 0 \in \mathbb{C}$ implies that $(1 - \lambda)|z|^2 = i|z|^4$, and hence there are no non-trivial solutions $(\lambda, z) \in \mathbb{R} \times \mathbb{C}$ of the equation $F(\lambda, z) = 0$. In particular, there are no bifurcation points. In this example, $\ker \partial_z F[(\lambda_0, 0)]$ is two-dimensional over $\mathbb{R}$. $\qquad \square$

8.2 LYAPUNOV-SCHMIDT REDUCTION

The Lyapunov-Schmidt procedure is a method for reducing the question of existence of solutions to an infinite-dimensional equation, locally in a neighbourhood of a known solution, to an equivalent one involving an equation in finite dimensions, quite commonly (though not always) in just two dimensions. Our setting is, as usual, the Banach spaces X and Y with a mapping $F \in C^k(U, Y)$ for some $k \in \mathbb{N}$, where U is open in $\mathbb{F} \times X$.

THEOREM 8.2.1 (Lyapunov-Schmidt Reduction) *Suppose*

$F(\lambda_0, x_0) = 0 \in Y$ *where* $(\lambda_0, x_0) \in U$,

the partial Fréchet derivative $L = \partial_x F[(\lambda_0, x_0)] : X \to Y$ *is a Fredholm operator,* $\ker(L) \neq \{0\}$ *and* $q \in \mathbb{N}$ *is the codimension of range* (L).

Then there exist two open sets, $U_0 \subset U$ *and* $V \subset \mathbb{F} \times \ker(L)$, *and two mappings,* $\psi \in C^k(V, X)$ *and* $h \in C^k(V, \mathbb{F}^q)$, *such that* $(\lambda_0, x_0) \in U_0$, $(\lambda_0, 0) \in V$ *and* $\psi(\lambda_0, 0) = x_0$ *with* $F(\lambda, x) = 0$ *and* $(\lambda, x) \in U_0$ *if and only if* $\psi(\lambda, \xi) = x$ *for some* $(\lambda, \xi) \in V$ *with* $h(\lambda, \xi) = 0$.

REMARKS 8.2.2 The infinite-dimensional problem $F(\lambda, x) = 0$ is "reduced" to the equivalent finite-dimensional problem "find $(\lambda, \xi) \in V \subset \mathbb{F} \times \ker(L)$ such that $h(\lambda, \xi) = 0$." In the event that F is $\mathbb{F}$-analytic it will be clear from the proof and the analytic implicit function theorem 4.5.4 that h and ψ will also be $\mathbb{F}$-analytic. $\qquad \square$

Proof. Since L is a Fredholm operator there exists a finite-dimensional (and therefore closed) subspace $Z \subset Y$ and a closed subspace $W \subset X$ such that

$$X = \ker(L) \oplus W \ \text{and} \ Y = Z \oplus \text{range}\,(L)$$

with dimension $Z = q$ (the codimension of the range of L). Hence there exists (see §2.6) a (bounded) projection $P : Y \to Y$ such that $\ker P$ is the range of L and the range of P is Z. In particular, $(I - P)Lx = Lx$ for all $x \in X$. Since $X = \ker(L) \oplus W$ it follows that $(I - P)L$ is a bijection, and hence a homeomorphism, from W onto range (L).

For $\xi \in \ker(L)$, $\eta \in W$ and $\lambda \in \mathbb{F}$ such that $(\lambda, x_0 + \xi + \eta) \in U$, let

$$G(\lambda, \xi, \eta) = (I - P)F(\lambda, x_0 + \xi + \eta).$$

Note that

$$G(\lambda_0, 0, 0) = (I - P)F(\lambda_0, x_0) = 0 \text{ and}$$
$$\partial_\eta G[(\lambda_0, 0, 0)]\eta = (I - P)\partial_x F[(\lambda_0, x_0)]\eta$$
$$= (I - P)L\eta \text{ for all } \eta \in W.$$

Hence $\partial_\eta G[(\lambda_0, 0, 0)]$, is a homeomorphism from W onto the range of L and, by the implicit function theorem 3.5.4 (4.5.4 when F is analytic), there exist open sets $U_0 \subset U$, $V \subset \mathbb{F} \times \ker(L)$,

a mapping $\phi \in C^k(V, W)$ such that $(\lambda_0, 0) \in V$, $(\lambda_0, x_0) \in U_0$,
$\phi(\lambda_0, 0) = 0$ and $G(\lambda, \xi, \phi(\lambda, \xi)) = 0$ for all $(\lambda, \xi) \in V$,

and such that

$$\{(\lambda, x_0 + \xi + \eta) \in U_0 : (I - P)F(\lambda, x_0 + \xi + \eta) = 0\}$$
$$= \{(\lambda, x_0 + \xi + \eta) : (\lambda, \xi) \in V \text{ and } \eta = \phi(\lambda, \xi)\}. \quad (8.1)$$

It therefore suffices to put

$$\psi(\lambda, \xi) = x_0 + \xi + \phi(\lambda, \xi) \text{ and } h(\lambda, \xi) = PF(\lambda, \psi(\lambda, \xi)) \in Z. \quad (8.2)$$

Then for all $(\lambda, \xi) \in V$,

$$h(\lambda, \xi) = 0 \text{ if and only if } PF(\lambda, x_0 + \xi + \phi(\lambda, \xi)) = 0$$
$$\text{if and only if } F(\lambda, x_0 + \xi + \phi(\lambda, \xi)) = 0,$$

because $(I - P)F(\lambda, x_0 + \xi + \phi(\lambda, \xi)) = 0$. Finally choose a basis for the q-dimensional space Z and thereby identify Z with $\mathbb{F}^q$. $\qquad\square$

REMARKS 8.2.3 In what follows we use the notation in (8.1) and (8.2), where P is the projection from $Z \oplus$ range (L) onto $Z \approx \mathbb{F}^q$. $\qquad\square$

8.3 CRANDALL-RABINOWITZ TRANSVERSALITY

In this section we present an important condition sufficient to guarantee that $(\lambda_0, 0)$ is a bifurcation point on the line of trivial solutions of $F(\lambda, x) = 0$. The following theory of bifurcation is due to Crandall and Rabinowitz [22] who were among the first to analyse such questions systematically using implicit function theorems.

THEOREM 8.3.1 *Suppose that X and Y are Banach spaces, that $F : \mathbb{F} \times X \to Y$ is of class C^k, $k \geq 2$, and that $F(\lambda, 0) = 0 \in Y$ for all $\lambda \in \mathbb{F}$. Suppose also that*

$L = \partial_x F[(\lambda_0, 0)]$ *is a Fredholm operator of index zero;*

$\ker(L)$ *is one-dimensional;*

$\ker(L) = \{\xi \in X : \xi = s\xi_0 \text{ for some } s \in \mathbb{F}\}, \ \xi_0 \in X \setminus \{0\};$

the transversality condition holds:

$$\partial^2_{\lambda, x} F[(\lambda_0, 0)](1, \xi_0) \notin range\,(L). \tag{8.3}$$

Then $(\lambda_0, 0)$ is a bifurcation point. More precisely, there exists $\epsilon > 0$ and a branch of solutions

$$\{(\lambda, x) = (\Lambda(s), s\chi(s)) : s \in \mathbb{F}, \ |s| < \epsilon\} \subset \mathbb{F} \times X,$$

such that $\Lambda(0) = \lambda_0$; $\chi(0) = \xi_0$;

$F(\Lambda(s), s\chi(s)) = 0$ *for all s with $|s| < \epsilon$;*
Λ *and $s \mapsto s\chi(s)$ are of class C^{k-1}, and χ is of class C^{k-2}, on $(-\epsilon, \epsilon)$;*
there exists an open set $U_0 \subset \mathbb{F} \times X$ such that $(\lambda_0, 0) \in U_0$ and

$$\{(\lambda, x) \in U_0 : F(\lambda, x) = 0, \ x \neq 0\}$$
$$= \{(\Lambda(s), s\chi(s)) : 0 < |s| < \epsilon\};$$

if F is analytic, χ and Λ are analytic functions on $(-\epsilon, \epsilon)$.

REMARKS 8.3.2 χ is a function from $(-\epsilon, \epsilon) \to X$. The notation in (8.3) is defined in Remark 3.2.4 where

$$\partial_{\lambda, x} F[(\lambda_0, 0)](1, \xi_0) = \lim_{t \to 0} \frac{\partial_x F[(\lambda_0 + t, 0)]\xi_0 - \partial_x F[(\lambda_0, 0)]\xi_0}{t} \in Y.$$

□

Proof. Let U_0, V, ϕ, ψ, and h be given by Lyapunov-Schmidt reduction of the equation $F(\lambda, x) = 0$ in a neighbourhood of the point $(\lambda_0, 0) \in \mathbb{F} \times X$. Note that (8.1) with $x_0 = 0$ implies that $\phi(\lambda, 0) = 0$, since $F(\lambda, 0) = 0$, for all $\lambda \in \mathbb{R}$. It is now easily confirmed that

$h(\lambda, 0) = PF(\lambda, 0 + \phi(\lambda, 0)) = 0$ for all $(\lambda, 0) \in V$;

$\partial_\xi h[(\lambda_0, 0)] = PL(I_\xi + \partial_\xi \phi[(\lambda, 0)]) = 0$ (I_ξ is the identity on $\ker(L)$);

$\partial^2_{\lambda, \xi} h[(\lambda_0, 0)](1, \xi_0) = P\partial^2_{\lambda, x} F[(\lambda_0, 0)](1, \xi_0) \neq 0;$

$h \in C^k(V, \mathbb{F})$, $k \geq 2$, where the one-dimensional space Z has, as in §8.2, been identified with $\mathbb{F}$ and $V \subset \mathbb{F} \times \ker(L)$ is an open set containing $(\lambda_0, 0)$.

Now define $g : V \to \mathbb{F}$ by

$$g(\lambda, \xi) = \int_0^1 \partial_\xi h[(\lambda, t\xi)] \, \xi_0 \, dt.$$

It is immediate that g is of class C^{k-1} (when F (and therefore h) is analytic, g is analytic) and that

$$g(\lambda_0, 0) = 0, \quad \partial_\lambda g(\lambda_0, 0) = \partial^2_{\lambda, \xi} h[(\lambda_0, 0)](1, \xi_0) \neq 0.$$

The implicit function theorem 3.5.4 now gives the existence of a mapping $\Lambda \in C^{k-1}(\{s \in \mathbb{F} : |s| < \epsilon\}, \mathbb{F})$ for some $\epsilon > 0$, such that $\Lambda(0) = \lambda_0$ and $g(\Lambda(s), s\xi_0) = 0$ if $|s| < \epsilon$. To complete the proof observe, since $h(\lambda, 0) = 0$, that

$$g(\lambda, \xi) = \begin{cases} \dfrac{h(\lambda, \xi)}{s} & \text{if } s \neq 0, \\[2mm] \partial_\xi h(\lambda, 0)\xi_0 & \text{if } s = 0, \end{cases} \quad \text{for } (\lambda, \xi) = (\lambda, s\xi_0) \in V.$$

Now put

$$\chi(s) = s^{-1}\psi(\Lambda(s), s\xi_0) \text{ for } 0 < |s| < \epsilon \text{ and } \chi(0) = \xi_0.$$

In fact

$$\lim_{0 \neq s \to 0} \chi(s) = \partial_\lambda \psi[(\lambda_0, 0)]\Lambda'(0) + \partial_\xi \psi[(\lambda_0, 0)]\xi_0 = \xi_0,$$

and so χ is continuous at $s = 0$. Indeed, for all s with $|s| < \epsilon$,

$$\chi(s) = \int_0^1 \partial_\lambda \psi[(\Lambda(ts), ts\xi_0)]\Lambda'(ts) + \partial_\xi \psi[(\Lambda(ts), ts\xi_0)]\xi_0 dt,$$

from which it follows easily that χ is of class C^{k-2}, and $\mathbb{F}$-analytic in the case that F, and consequently Λ and ψ, are $\mathbb{F}$-analytic. This completes the proof. $\square$

EXAMPLES 8.3.3 **(Concerning Transversality)** (a) Let $\mathbb{F} = \mathbb{R}$, $X = Y = \mathbb{R}$ and let $F(\lambda, x) = x(\lambda^2 + x^2)$. Clearly $F : \mathbb{R} \times \mathbb{R} \to \mathbb{R}$ is C^2, $F(\lambda, 0) = 0$ for all $\lambda \in \mathbb{R}$, $L = \partial_x F[(0, 0)]$ is the zero operator, $\ker(L) = X = \mathbb{R}$ and $R(L) = \{0\}$, but the transversality condition is not satisfied because $\partial^2_{\lambda, x} F[(0, 0)]$ is the zero element of $\mathcal{M}^2(\mathbb{R}; \mathbb{R})$. It is easily seen that $(0, 0) \in \mathbb{R} \times X$ is not a bifurcation point.

(b) If, in example (a), $F(\lambda, x) = x(\lambda + x^2)$, then the transversality condition holds and (of course) $(0, 0)$ is a bifurcation point.

(c) If however $F(\lambda, x) = x(\lambda^3 + x^2)$, then $(0, 0)$ is a bifurcation point although the transversality condition fails. $\square$

PROPOSITION 8.3.4 *(a) Let the hypotheses of Theorem 8.2.1 be satisfied and let U_0, V, h and ψ be given by its conclusion. Then U_0 and V can be chosen sufficiently small that for all $(\lambda, \xi) \in V$*

$$\dim \ker \left(\partial_x F[(\lambda, \psi(\lambda, \xi))] \right) = \dim \left(\ker(\partial_\xi h[(\lambda, \xi)]) \right).$$

(b) Suppose that the hypotheses of Theorem 8.3.1 are satisfied and that, in part (a) above, $U = \mathbb{F} \times X$ and $(\lambda_0, x_0) = (\lambda_0, 0)$. Now let Λ, χ, ϵ be as in the conclusion of Theorem 8.3.1. Then

$$\dim \ker \left(\partial_x F[(\Lambda(s), s\chi(s))] \right) \in \{0, 1\}$$

and, for s with $0 < |s| < \epsilon$, $\dim \ker \left(\partial_x F[(\Lambda(s), s\chi(s))] \right) = 1$ if and only if $\Lambda'(s) = 0$.

Proof. From the identity

$$(I - P)F(\lambda, x_0 + \xi + \phi(\lambda, \xi)) \equiv 0, \quad (\lambda, \xi) \in V,$$

it follows that

$$(I - P)\partial_x F[(\lambda, x_0 + \xi + \phi(\lambda, \xi))](v + \partial_\xi \phi[(\lambda, \xi)]v) \equiv 0$$

for all $v \in \ker(L)$. Therefore, if $\partial_x F[(\lambda, x_0 + \xi + \phi(\lambda, \xi))](v + w) = 0$ with $v \in \ker(L)$ and $w \in W$, then $w - \partial_\xi \phi[(\lambda, \xi)]v \in W$ and

$$(I - P)\partial_x F[(\lambda, x_0 + \xi + \phi(\lambda, \xi))](w - \partial_\xi \phi[(\lambda, \xi)]v) = 0.$$

Since $(I - P)L : W \to \text{range}\,(L)$ is bijective, $(I - P)\partial_x F[(\lambda, x_0 + \xi + \phi(\lambda, \xi))]\big|_W$ is also a bijection (in particular it is injective) when the sets U_0 and V have been chosen with sufficiently small diameters (see §2.5). Therefore $w = \partial_\xi \phi[(\lambda, \xi)]v$ and so

$$\partial_x F[(\lambda, x_0 + \xi + \phi(\lambda, \xi))](v + w) = 0 \text{ for some } w \in W$$
$$\text{if and only if } P\partial_x F[(\lambda, x_0 + \xi + \phi(\lambda, \xi))](v + \partial_\xi \phi[(\lambda, \xi)]v) = 0$$
$$\text{if and only if } \partial_\xi h[(\lambda, \xi)]v = 0.$$

Part (a) follows. In part (b), $\dim \ker \left(\partial_x F[(\Lambda(s), s\chi(s))] \right) \in \{0, 1\}$ because $\partial_\xi h[(\lambda, \xi)]$ maps the one-dimensional space span ξ_0 into itself. For the second part of (b) note that, for s with $|s| < \epsilon$,

$$0 = \frac{d}{ds} h(\Lambda(s), s\xi_0) = \partial_\lambda h[(\Lambda(s), s\xi_0)]\Lambda'(s) + \partial_\xi h[(\Lambda(s), s\xi_0)]\xi_0$$
$$= s\partial_\lambda g[(\Lambda(s), s\xi_0)]\Lambda'(s) + \partial_\xi h[(\Lambda(s), s\xi_0)]\xi_0$$

with $\partial_\lambda g[(\Lambda(s), s\xi_0)] \neq 0$ if $\epsilon > 0$ is sufficiently small. (Recall the hypothesis that

$$\partial_\lambda g[(\lambda_0, 0)](1) = \partial^2_{\lambda, \xi} h[(\lambda_0, 0)](1, \xi_0) = P\partial^2_{\lambda, x} F[(\lambda_0, 0)](1, \xi_0) \neq 0.)$$

This proves that, for any s with $0 < |s| < \epsilon$,

$$\dim \ker \left(\partial_x F[(\Lambda(s), s\chi(s))] \right) = 1 \text{ if and only if } \Lambda'(s) = 0.$$

$\square$

8.4 BIFURCATION FROM A SIMPLE EIGENVALUE

Here is a case where transversality is trivial to verify. Recall the definition of a simple eigenvalue 2.7.8

THEOREM 8.4.1 **(Bifurcation from a Simple Eigenvalue)** *Suppose that the real Banach space X is continuously embedded in the real Banach space Y and that $\{(\lambda, 0) : \lambda \in \mathbb{R}\} \subset U \subset \mathbb{R} \times X$, where U is open. Suppose that $F \in C^k(U, Y)$, $k \geq 2$ and for all $\lambda \in \mathbb{R}$,*

$$F(\lambda, 0) = 0 \quad \text{and} \quad \partial_x F[(\lambda, 0)] = \lambda \iota - A, \tag{8.4}$$

where ι is the continuous embedding of X in Y. Then every simple eigenvalue λ_0 of A is a bifurcation point and the conclusion of Theorem 8.3.1 holds.

Proof. Because of the hypotheses, the transversality condition (8.3) demands that $\lambda_0 \iota - A$ be Fredholm of index 0 with one-dimensional kernel spanned by ξ_0 say, and that

$$\xi_0 = \partial^2_{\lambda,x} F[(\lambda_0, 0)](1, \xi_0) \notin \text{range}\,(\lambda_0 \iota - A).$$

However, this is precisely the requirement that λ_0 is a simple eigenvalue of A. The result now follows as a special case of Theorem 8.3.1. $\qquad\qquad\square$

Suppose the hypotheses of Theorem 8.4.1 hold. Let $y^* \in Y^*$ be such that $y^*(\xi_0) = \|\xi_0\|$, $\|y^*\| = 1$, where $y^*(\text{range}\,(\lambda_0 \iota - A)) = 0$ and ξ_0 spans the kernel of $\lambda_0 \iota - A$, as in the proof of Proposition 3.6.1.

Let $(\Lambda(s), s\chi(s))$ be as in the conclusion of Theorem 8.4.1 and let

$$L(s) = \partial_x F[(\Lambda(s), s\chi(s))] \in \mathcal{L}(X, Y), \quad s \in (-\epsilon, \epsilon),$$

$L(0) = \lambda_0 \iota - A$. Then, by Theorem 3.6.1, there is a C^{k-1}-curve $\{(\mu(s), \xi(s)) : s \in (-\epsilon, \epsilon)\} \subset \mathbb{R} \times X$ such that $(\mu(0), \xi(0)) = (0, \xi_0)$,

$$L(s)\xi(s) = \mu(s)\iota\xi(s) \quad \text{and} \quad y^*(\iota\xi(s)) = 1, \quad s \in (-\epsilon, \epsilon),$$

where $\xi(s) = \xi_0 + \eta(s)$, η is of class C^{k-1}, $\eta(0) = 0$ and $\iota\eta(s) \in X \cap \text{range}\,(L(0))$.

The next result explains the direction in which $\mu(s)$ moves as s passes through 0, an observation which is important when deciding the stability and the Morse index of the bifurcating solutions, see §11.3.

PROPOSITION 8.4.2 *In this notation,*

$$\lim_{\substack{s \to 0 \\ s\Lambda'(s) \neq 0}} \frac{\mu(s)}{s\Lambda'(s)} = -1.$$

Proof. For $s \in (-\epsilon, \epsilon)$, in the notation of (8.2) let

$$x(s) = s\chi(s) = \psi(\Lambda(s), s\xi_0) = s\xi_0 + \phi(\Lambda(s), s\xi_0) = s\xi_0 + \rho(s), \text{ say.}$$

(This is the definition of ρ.) Since $F(\Lambda(s), x(s)) = 0$, differentiation gives

$$\partial_\lambda F[(\Lambda(s), x(s)]\Lambda'(s) + L(s)x'(s) = 0$$

for all $s \in (-\epsilon, \epsilon)$. Therefore $L(s)\xi(s) = \mu(s)\iota\,\xi(s)$, $s \in (-\epsilon, \epsilon)$, implies that

$$L(s)\big(\xi(s) - x'(s)\big) = \partial_\lambda F[(\Lambda(s), x(s))]\Lambda'(s) + \mu(s)\iota\xi(s).$$

Since $\partial_{\lambda^2}^2 F[(\lambda_0, 0)] = 0$ and $\partial_{\lambda,x}^2 F[(\lambda_0, 0)](\lambda, x) = \lambda x$, we find that

$$\begin{aligned}
&\partial_\lambda F[(\Lambda(s), x(s))] \\
&= \partial_{\lambda^2}^2 F[(\lambda_0, 0)](\Lambda(s) - \lambda_0) \\
&\quad + \partial_{\lambda,x}^2 F[(\lambda_0, 0)]x(s) + o(|\Lambda(s) - \lambda_0|) + o(\|x(s)\|) \\
&= s\iota\xi_0 + \iota\rho(s) + o(|\Lambda(s) - \lambda_0|) + o(\|x(s)\|) = s\xi_0 + o(|s|)
\end{aligned}$$

as $s \to 0$. Therefore, since $\xi(s) - x'(s) = \eta(s) - \rho'(s)$,

$$L(s)\big(\eta(s) - \rho'(s)\big) = \Lambda'(s)(s\iota\xi_0 + o(|s|)) + \mu(s)\iota(\xi_0 + \eta(s)),$$

and so

$$\begin{aligned}
L(0)(\eta(s) - \rho'(s)) = {}&\Lambda'(s)(s\iota\xi_0 + o(|s|)) + \mu(s)\iota(\xi_0 + \eta(s)) \\
&+ \big(L(0) - L(s)\big)(\eta(s) - \rho'(s)). \quad (8.5)
\end{aligned}$$

By Lemma 2.7.9, $X_1 = X \cap \text{range}\,(L(0))$ is closed in X and $L(0) = \lambda_0\iota - A$ is a homeomorphism from X_1 onto range $(L(0)) \subset Y$. Since $\eta(s) - \rho'(s) \in X_1$

$$\|\eta(s) - \rho'(s)\|_X \le \text{const.}\,(|s\Lambda'(s)| + |\mu(s)|), \quad s \in (-\epsilon, \epsilon).$$

Applying y^* to both sides of (8.5) gives

$$-\big(s\Lambda'(s) + \mu(s)\big)\|\xi_0\| = o(|s|)\Lambda'(s) + y^*\big((L(0) - L(s))(\eta(s) - \rho'(s))\big).$$

Therefore, for $\epsilon > 0$ sufficiently small,

$$|s\Lambda'(s) + \mu(s)| \le o(|s|)|\Lambda'(s)| + o(1)|\mu(s)|, \quad s \in (-\epsilon, \epsilon). \quad (8.6)$$

Since η is continuous and $\eta(0) = 0$,

$$\begin{aligned}
|\mu(s)| &\le |s\Lambda'(s)| + |s\Lambda'(s) + \mu(s)| \\
&\le |s\Lambda'(s)| + o(|s|)\,|\Lambda'(s)| + o(1)|\mu(s)| \text{ as } s \to 0.
\end{aligned}$$

Therefore

$$|\mu(s)| \le \text{const.}\,|s\Lambda'(s)|, \quad s \in (-\epsilon, \epsilon).$$

Similarly, $|s\Lambda'(s)| \le \text{const.}\,|\mu(s)|$. Therefore (8.6) implies that

$$|s\Lambda'(s) + \mu(s)| = o(|s\Lambda'(s)|) \text{ and } |s\Lambda'(s) + \mu(s)| = o(|\mu(s)|)$$

as $s \to 0$. The result now follows.

$\square$

We now give two applications of the local-bifurcation method just explained. The first was featured in the Introduction where a global theory was developed by solving the equation more-or-less explicitly. Here we see that the bifurcating of solutions observed there is a consequence of general abstract considerations and the theory of bifurcation from a simple eigenvalue.

8.5 BENDING AN ELASTIC ROD II

We now show how local bifurcation theory applies to the boundary-value problem (1.3):

$$\phi''(x) + \lambda \sin \phi(x) = 0 \text{ for } x \in [0, L], \quad \phi'(0) = \phi'(L) = 0, \qquad (8.7)$$

where L is fixed and $\lambda > 0$ is the parameter in the problem. Let

$$F = \mathbb{R}, \quad X = \{\phi \in C^2[0, L] : \phi'(0) = \phi'(L) = 0\}, \quad Y = C[0, L],$$

and define $F(\lambda, \phi) = \phi'' + \lambda \sin \phi$. It is not difficult to see (as in Example 4.3.6) that $F : \mathbb{R} \times X \to Y$ is $\mathbb{R}$-analytic. Moreover

$$\partial_\phi F[(\lambda, 0)]\phi = 0$$

if and only if

$$\phi'' + \lambda \phi = 0 \in Y \text{ and } \phi'(0) = \phi'(L) = 0.$$

This linear boundary-value problem has solutions (λ, ϕ) with $\lambda > 0$ and $\phi \neq 0$ if and only if $\lambda \in \{\lambda_K = \left(\frac{K\pi}{L}\right)^2 : K \in \mathbb{N}\}$. Then ϕ is a multiple of ϕ_K where $\phi_K(s) = \cos \frac{K\pi s}{L}$ and $\dim \ker \partial_\phi F[(\lambda_K, 0)] = 1$. Moreover

$$\text{range} \left(\partial_\phi F[(\lambda_K, 0)]\right) = \{v \in C[0, L] : \int_0^L v(s)\phi_K(s)ds = 0\}.$$

To see this note that if $u \in X$ and $u'' + \lambda_K u = v \in Y$, then an integration by parts implies that

$$\int_0^L v(s)\phi_K(s)ds = \int_0^L (u''(s) + \lambda_n u(s))\phi_K(s)ds$$

$$= \int_0^L (\phi_K''(s) + \lambda_K \phi(s))u(s)ds = 0.$$

That every $v \in Y$ with

$$\int_0^L v(s)\phi_K(s)ds = 0$$

is in the range of $\partial_\phi F[(\lambda_K, 0)]$ can be shown using the variation of constants formula.

Let $A : X \to Y$ be defined by $A\phi = -\phi''$. Then λ_K, $K \in \mathbb{N}$, is a simple eigenvalue of A. Therefore, for all $K \in \mathbb{N}$, there is a bifurcation from a simple eigenvalue for (8.7) at every point $(\lambda_K, 0)$ on the line of trivial solutions.

8.6 BIFURCATION OF PERIODIC SOLUTIONS

This section concerns a simple example of the existence, via bifurcation from a line of trivial solutions, of non-constant but periodic, solutions of a differential equation.

Let $\delta > 0$ and suppose that the functions A, $B \in C^2\big((-\delta, \delta) \times \mathbb{R} \times (-\delta, \delta); \mathbb{R}\big)$ are 2π-periodic in the second variable, t (time). Now consider the boundary-value problem on $[0, 2\pi]$

$$\dot{w}(t) + \lambda w(t) + w(t)^2 A(w(t), t, \lambda) + \lambda^2 w(t) B(w(t), t, \lambda) = 0,$$
$$w(0) = w(2\pi), \quad w \in C^1[0, 2\pi],$$

where $\dot{w} = dw/dt$. Note that for all $\lambda \in \mathbb{R}$, $w = 0$ is a time-independent solution of this problem.

To establish the existence of time-dependent solutions let

$$\mathbb{F} = \mathbb{R}, \quad X = \{u \in C^1[0, 2\pi] : u(0) = u(2\pi), u'(0) = u'(2\pi)\},$$
$$Y = \{v \in C[0, 2\pi] : v(0) = v(2\pi)\}$$

and, for $0 \le t \le 2\pi$ and $u \in X$, let

$$F(\lambda, u)(t) = \dot{u}(t) + \lambda u(t) + u(t)^2 A(u(t), t, \lambda) + \lambda^2 u(t) B(u(t), t, \lambda).$$

Then $F \in C^2(U, Y)$, where

$$U = \{(\lambda, u) : |\lambda| < \delta, \max_{0 \le t \le 2\pi} |u(t)| < \delta\} \subset \mathbb{R} \times X.$$

Moreover

$$\partial_u F[(0, 0)]u = \dot{u}, \, u \in X,$$

$$\ker\big(\partial_u F[(0, 0)]\big) = \{u \in X : u \text{ is a constant}\},$$

$$\text{range}\big(\partial_u F[(0, 0)]\big) = \{v \in Y : \int_0^{2\pi} v(t)dt = 0\},$$

and therefore $\partial_u F[(0, 0)]$ is a Fredholm operator of index zero. Let $\xi_0 = 1$. It follows that

$$\partial^2_{\lambda, u} F[(0, 0)](1, \xi_0) = \xi_0 \notin \text{range}\big(\partial_u F[(0, 0)]\big).$$

Therefore, by Theorem 8.3.1, there exists a C^1-curve

$$\{(\Lambda(s), s\chi(s)) \in U : s \in (-\epsilon, \epsilon)\}$$

such that $F(\Lambda(s), s\chi(s)) = 0$, $\chi(0) = \xi_0$ and $\Lambda(0) = 0$.

For $s \in (0, \epsilon)$ sufficiently small, $(\lambda, w) = (\Lambda(s), s\chi(s))$ is a periodic solution of our problem for which w is positive (because $\chi(s) = 1 + \eta(s)$ and $\eta(0) = 0$ in X) on $[0, 2\pi]$. To ensure that, for $\epsilon > 0$ sufficiently small, this is not a constant solution we need an extra hypothesis, such as

$$\partial_t A[(0, t, 0)] \neq 0, \tag{8.8}$$

which ensures that constant solutions do not bifurcate from the line of trivial solutions at $(0, 0)$. Suppose that (8.8) holds and that there is a sequence $\{(\lambda_n, c_n)\}$ of solutions, where $c_n \neq 0$ is a constant, converging to $(0, 0)$. Then

$$\lambda_n + c_n A(c_n, t, \lambda_n) + \lambda_n^2 B(c_n, t, \lambda_n) = 0.$$

After dividing by $\|(\lambda_n, c_n)\|$ and taking the limit of a subsequence as $n \to \infty$, we obtain a vector (λ^*, c^*) with $\|(\lambda^*, c^*)\| = 1$ satisfying $\lambda^* + c^* A(0, t, 0) = 0$. This contradicts (8.8).

8.7 NOTES ON SOURCES

This material is now completely standard, see Ambrosetti and Prodi [2], Chow and Hale [19], Crandall and Rabinowitz [22, 23], Schwartz [52], Stuart [54, 55], Zeidler [68], Toland [61], and many other references.

Chapter Nine

Global Bifurcation Theory

Let X, Y be Banach spaces over $\mathbb{R}$, let $U \subset \mathbb{R} \times X$ and $F : U \to Y$ be an $\mathbb{R}$-analytic function. Suppose that

(G1) $(\lambda, 0) \in U$ and $F(\lambda, 0) = 0$ for all $\lambda \in \mathbb{R}$.

(G2) $\partial_x F[(\lambda, x)]$ is a Fredholm operator of index zero when $F(\lambda, x) = 0$, $(\lambda, x) \in U$.

(G3) For some $\lambda_0 \in \mathbb{R}$,

$$\ker \left(\partial_x F[(\lambda_0, 0)] \right) = \{ s\xi_0 : s \in \mathbb{R} \},$$
$$\partial^2_{\lambda, x} F[(\lambda_0, 0)](1, \xi_0) \notin \operatorname{range} \left(\partial_x F[(\lambda_0, 0)] \right).$$

By Theorem 8.3.1, there exists an analytic function $(\Lambda, \kappa) : (-\epsilon, \epsilon) \to \mathbb{R} \times X$ such that $(\lambda, x) = (\Lambda(s), \kappa(s))$ satisfies $F(\lambda, x) = 0$ for all $s \in (-\epsilon, \epsilon)$, $\Lambda(0) = \lambda_0$ and $\kappa'(0) = \xi_0$. (In the notation of §8.3, $\kappa(s) = s\chi(s)$.) Let

$$\mathcal{R}^+ = \{ (\Lambda(s), \kappa(s)) : s \in (0, \epsilon) \},$$
$$\mathcal{S} = \{ (\lambda, x) \in U : F(\lambda, x) = 0 \},$$
$$\mathcal{T} = \{ (\lambda, x) \in \mathcal{S} : x \neq 0 \}.$$

Suppose $\epsilon > 0$ is sufficiently small that $\kappa'(s) \neq 0$ for $s \in (-\epsilon, \epsilon)$ and $\mathcal{R}^+ \subset \mathcal{T}$.

9.1 GLOBAL ONE-DIMENSIONAL BRANCHES

The following result gives a global extension of the function (Λ, κ) from $(0, \epsilon)$ to $(0, \infty)$ in the $\mathbb{R}$-analytic case. However, in proving it we develop a general approach to the global extendability of one-parameter curves of non-singular solutions to an equation $F(\lambda, x) = 0$.

THEOREM 9.1.1 *Suppose (G1)–(G3) hold, $\Lambda' \not\equiv 0$ on $(-\epsilon, \epsilon)$ and that in $\mathbb{R} \times X$ all bounded closed subsets of $\mathcal{S}$ are compact. Then there exists a continuous curve $\mathfrak{R}$ which extends $\mathcal{R}^+$ as follows.*

(a) $\mathfrak{R} = \{ (\Lambda(s), \kappa(s)) : s \in [0, \infty) \} \subset U$ *where* $(\Lambda, \kappa) : [0, \infty) \to \mathbb{R} \times X$ *is continuous.*

(b) $\mathcal{R}^+ \subset \mathfrak{R} \subset \mathcal{S}$.

(c) *The set* $\{s \geq 0 : \ker\left(\partial_x F[(\Lambda(s), \kappa(s))]\right) \neq \{0\}\}$ *has no accumulation points.*

(d) *At each point, $\mathfrak{R}$ has a local analytic re-parameterization in the following sense. In a right neighbourhood of $s = 0$, $\mathfrak{R}$ and $\mathcal{R}^+$ coincide. For each $s^* \in (0, \infty)$ there exists $\rho^* : (-1, 1) \to \mathbb{R}$ which is continuous, injective, and*

$$\rho^*(0) = s^*, \quad t \mapsto (\Lambda(\rho^*(t)), \kappa(\rho^*(t))), \quad t \in (-1, 1), \text{ is analytic.}$$

Furthermore Λ is injective on a right neighbourhood of 0 and for $s^* > 0$ there exists $\epsilon^* > 0$ such that Λ is injective on $[s^*, s^* + \epsilon^*]$ and on $[s^* - \epsilon^*, s^*]$.

(e) *One of the following occurs.*

    (i) $\|(\Lambda(s), \kappa(s))\| \to \infty$ *as $s \to \infty$.*

    (ii) $(\Lambda(s), \kappa(s))$ *approaches the boundary of U as s tend to ∞.*

    (iii) $\mathfrak{R}$ *is a closed loop. In other words, for some $T > 0$,*

$$\mathfrak{R} = \{(\Lambda(s), \kappa(s)) : 0 \leq s \leq T\}$$

and $(\Lambda(T), \kappa(T)) = (\lambda_0, 0)$. We may suppose that $T > 0$ is the smallest such T and

$$(\lambda(s + T), \kappa(s + T)) = (\Lambda(s), \kappa(s)) \text{ for all } s \geq 0.$$

(f) *If, for some $s_1 \neq s_2$,*

$$(\Lambda(s_1), \kappa(s_1)) = (\Lambda(s_2), \kappa(s_2)) \text{ where } \ker \partial_x F[(\Lambda(s_1), \kappa(s_1))] = \{0\},$$

then (e)(iii) occurs and $|s_1 - s_2|$ is an integer multiple of T.

In particular, $(\Lambda, \kappa) : [0, \infty) \to \mathcal{S}$ is locally injective.

REMARKS 9.1.2 (1) There is no claim that $\mathfrak{R}$ is a maximal connected subset of $\mathcal{S}$. Other curves or manifolds in $\mathcal{S}$ may intersect $\mathfrak{R}$.

(2) $\mathfrak{R}$ may self-intersect in the sense that while $s \mapsto (\Lambda(s), \kappa(s))$ is locally injective, it need not be globally injective. For example, in part (e)(iii) of the theorem it is clearly not globally injective.

(3) In part (d) it can happen that the parametrization has zero derivative, in which case $\{(\Lambda(s), \kappa(s)) : |s - s^*| < \delta^*\} \subset \mathfrak{R}$ may not be a smooth curve even though it has a local analytic parameterization at every point. Of course, for δ^* sufficiently small, the two segments of the set $\{(\Lambda(s), \kappa(s)) : 0 < |s - s^*| < \delta^*\}$, with $(\Lambda(s^*), \kappa(s^*))$ deleted, are smooth and can be parameterized by λ.

(4) Alternative (e)(i) is much stronger than the claim that $\mathfrak{R}$ is unbounded in $\mathbb{R} \times X$.   □

Proof. The proof of Theorem 9.1.1 is organized below in a few short steps. Let

$$\mathfrak{N} = \{(\lambda, x) \in \mathcal{S} : \ker\left(\partial_x F[(\lambda, x)]\right) = \{0\}\}$$

(the non-singular solutions of $F(\lambda, x) = 0$). By hypothesis $\Lambda \not\equiv 0$ on $(-\epsilon, \epsilon)$ and Λ is $\mathbb{R}$-analytic. Therefore Λ' is nowhere zero on $(-\epsilon, \epsilon) \setminus \{0\}$ for $\epsilon > 0$ sufficiently small and, by Proposition 8.3.4(b), we may assume that $\epsilon > 0$ is such that

$$\mathcal{R}^+ \subset \mathfrak{N}. \tag{9.1}$$

DEFINITION 9.1.3 (Distinguished Arcs) *A distinguished arc is a maximal connected subset of $\mathfrak{N}$.*

Hypothesis $G(2)$ and the analytic implicit function Theorem 4.5.4 ensure that a distinguished arc $\mathcal{I}$ is the graph of an $\mathbb{R}$-analytic function of λ. More precisely, if $\mathcal{I}$ is a distinguished arc then there exists a (possibly infinite) open interval I and an $\mathbb{R}$-analytic function $g : I \to X$ such that

$$\{(\lambda, g(\lambda)) : \lambda \in I\} = \mathcal{I}. \tag{9.2}$$

Step 1. (Lyapunov-Schmidt Reduction) We need to connect the theory of varieties with the present infinite-dimensional problem. To study the structure of $\mathcal{S}$ in a neighbourhood of a point $(\lambda_*, x_*) \in \mathcal{S} \setminus \mathfrak{N}$ we use the Lyapunov-Schmidt procedure. Theorem 8.2.1 yields the existence of
a neighbourhood V of $(\lambda_*, 0)$ in $\mathbb{R} \times \ker(\partial_x F[(\lambda_*, x_*)])$,
$\mathbb{R}$-analytic maps $\psi : V \to X$,
$h : V \to \mathbb{R}^q$ ($q = \dim \ker(\partial_x F[(\lambda_*, x_*)])$) with

(a) $\psi(\lambda_*, 0) = x_*$ and $(\lambda, \psi(\lambda, \xi)) \in U$ for $(\lambda, \xi) \in V$.

(b) For all $(\lambda, \xi) \in V$, $h(\lambda, \xi) = 0$ if and only if $F(\lambda, \psi(\lambda, \xi)) = 0$.

(c) If $F(\lambda, x) = 0$, $(\lambda, x) \in U$ and $\|(\lambda, x) - (\lambda_*, x_*)\|$ is sufficiently small, then there exists $\xi \in \ker(\partial_x F[(\lambda_*, x_*)])$ such that $(\lambda, \xi) \in V$ and $x = \psi(\lambda, \xi)$.

(d) $\dim \ker(\partial_x F[(\lambda, \psi(\lambda, \xi))]) = \dim \ker(\partial_\xi h[(\lambda, \xi)])$, $(\lambda, \xi) \in V$.

Recall the notation of §7. The analytic function $h : V \to \mathbb{R}^q$ may be identified with the set of its q component functions each of which maps V into $\mathbb{R}$ analytically. Therefore we may define an $\mathbb{R}$-analytic variety A and a manifold M by

$$A = \mathrm{var}\,(V, \{h\}) = \{(\lambda, \xi) \in V : h(\lambda, \xi) = 0\},$$
$$M = \{(\lambda, \xi) \in V : (\lambda, \psi(\lambda, \xi)) \in \mathfrak{N}\}.$$

By Proposition 8.3.4(a), the elements of M are 1-regular points of A. Let $\{M_j : j \in J\}$ denote those non-empty connected components of M which have the property that $\gamma_{(\lambda_*, 0)}(M_j) \neq \emptyset$.

Since h is an $\mathbb{R}$-analytic function on the $(q+1)$-dimensional real vector space V, the q components of $h(\lambda, \xi)$ are real functions defined locally in a neighbourhood of $(\lambda_*, 0) \in V$ by a Taylor series, the nth term of which is a sum of terms of the form

$$h^*_{k_1,\cdots,k_{q+1}} x_1^{k_1} \cdots x_{q+1}^{k_{q+1}},$$

where $k_1 + \cdots + k_{q+1} = n$ and $h^*_{k_1,\cdots,k_{q+1}} \in \mathbb{R}$.

Here $(x_1, \cdots x_{q+1}) \in \mathbb{R}^{q+1}$ are the coefficients of $(\lambda_*, 0) - (\lambda, \xi)$ in some linear coordinate system. Replacing $(x_1, \cdots, x_{q+1}) \in \mathbb{R}^{q+1}$ with $(z_1, \cdots, z_{q+1}) \in \mathbb{C}^{q+1}$ leads to a real-on-real $\mathbb{C}$-analytic extension h^c of h in a complex neighbourhood V^c of $(\lambda_*, 0)$ and a corresponding $\mathbb{C}$-analytic variety. Let

$$A^c = \mathrm{var}\,(V^c, \{h^c\}) = \{(\lambda, \xi) \in V^c : h^c(\lambda, \xi) = 0\},$$
$$M^c = \{(\lambda, \xi) \in V^c : \ker(\partial_\xi h^c[(\lambda, \xi)]) = \{0\}\},$$

and let $\{M^c_j : j \in J^c\}$ be the non-empty connected components of M^c with $\gamma_{(\lambda_*,0)}(\mathbb{R}^{q+1} \cap M^c_j) \neq \emptyset$. Note that for each $j \in J$ there exists $\hat{j} \in J^c$ such that $M_j \subset M^c_{\hat{j}}$.

Step 2. (Application of the Structure Theorem) Theorem 7.4.7 (d)–(f) on the structure of complex analytic varieties, when applied to A^c gives, for each $j \in J^c$, the existence of a real-on-real branch B_j with

$$\gamma_{(\lambda_*,0)}(M^c_j) \subset \gamma_{(\lambda_*,0)}(\overline{B}_j), \quad \dim B_j = 1 \text{ and } B_j \subset A^c.$$

By making the neighbourhood V^c smaller if necessary, we may suppose that $B_j \setminus \{(\lambda_*, 0)\} \subset M^c_j$. By Theorem 7.4.7 there are finitely many branches and hence finitely many M^c_j and M_j. By Theorem 7.5.1 each of these one-dimensional branches B_j admits a $\mathbb{C}$-analytic parameterization in a neighbourhood of $(\lambda_*, 0)$.

We now return to the setting of $\mathbb{R}^n$. From Corollary 7.5.3, we obtain that $\overline{M}$, locally near $(\lambda_*, 0)$, is the union of a finite number of curves which *pass through* $(\lambda_*, 0)$ in V, intersect one another only at $(\lambda_*, 0)$ and are given by the parameterization (7.16). Thus, in our previous notation, each $M_j, j \in J$, is paired, *in a unique way* with *another* $M_{\bar{j}}, \bar{j} \in J$, so that their union with the point $(\lambda_*, 0)$ forms one of these curves in V.

This observation can be lifted to infinite dimensions as follows. Suppose that $\mathcal{I}$ in (9.2) is a distinguished arc where $I = (a, b)$ with $(b, g(b)) \in \mathcal{S} \setminus \mathfrak{N}$. Then the germ $\gamma_{(\lambda_*,0)}(\mathcal{I})$ coincides with the germ of the image under the mapping $(\lambda, \xi) \mapsto (\lambda, \psi(\lambda, \xi))$ of M_j for some $j \in J$. Hence $\mathcal{I}$ has a unique extension beyond $(b, g(b))$ given by the image of $M_{\bar{j}}$ under the same mapping.

DEFINITION 9.1.4 (Routes of Length N) *A route of length $N \in \mathbb{N} \cup \{\infty\}$ is a set $\{\mathcal{A}_n : 0 \leq n < N\}$ of distinguished arcs and a set $\{(\lambda_n, x_n) : 0 \leq n < N\} \subset \mathbb{R} \times X$ such that*

(a) $(\lambda_0, x_0) = (\lambda_0, 0)$ is the bifurcation point;

(b) $\mathcal{R}^+ \subset \mathcal{A}_0$;

(c) For $N > 1$ and $0 \le n < N - 1$,

$$(\lambda_{n+1}, x_{n+1}) \in \left(\partial \mathcal{A}_n \cap \partial \mathcal{A}_{n+1}\right) \setminus \{(\lambda_n, x_n)\}$$

and there exists an injective $\mathbb{R}$-analytic map $\rho : (-1, 1) \to \mathcal{A}_n \cup \mathcal{A}_{n+1} \cup \{(\lambda_{n+1}, x_{n+1})\}$ with $\rho(0) = (\lambda_{n+1}, x_{n+1})$. Hence $\mathcal{A}_{n+1}$ is uniquely determined by $\mathcal{A}_n$ and vice versa.

(d) The mapping $n \mapsto \mathcal{A}_n$ is injective.

Step 3. (Existence of a Maximal Route) That $\{\mathcal{A}_0\}$, $\{(\lambda_0, 0)\}$ is a route of length 1 with $(\lambda_0, 0) \in \partial \mathcal{A}_0$ is obvious from the discussion leading to (9.1). Parts (c) and (d) of the definition of a route imply that $\mathcal{A}_{n+1} \ne \mathcal{A}_n$ and that $\mathcal{A}_{n+1}$ is uniquely determined by $\mathcal{A}_n$. Therefore if

$$\{\mathcal{A}_n^j, \ 0 \le n < N_j\}, \ \ \{(\lambda_n^j, x_n^j) : 0 \le n < N_j\}, \ \ j \in \{1, 2\},$$

are two routes with $N_1 \le N_2$ it follows that

$$\lambda_n^1 = \lambda_n^2, \ \ x_n^1 = x_n^2 \text{ for all } n \text{ with } 0 \le n < N_1.$$

Hence, under the hypotheses of Theorem 9.1.1, there exists a maximal route of length $N \in \mathbb{N} \cup \{\infty\}$ which we denote by

$$\{\mathcal{A}_n, (\lambda_n, x_n)\} : 0 \le n < N\}.$$

Step 4. (Parameterization of a Maximal Route) Because of the remark following Definition 9.1.3,

$$\mathcal{A}_n = \{(\Lambda_n(s), \kappa_n(s)), \ s \in (n, n+1)\}, \ \ 0 \le n < N,$$

where (Λ_n, κ_n) is an $\mathbb{R}$-analytic function. There are three cases: $N = \infty$; $N < \infty$ when $\overline{\mathcal{A}}_{N-1}$ is not a compact subset of U; and $N < \infty$ when $\overline{\mathcal{A}}_{N-1}$ is a compact subset of U.

Suppose that $\overline{\mathcal{A}}_n$ is a bounded closed subset of U for some $0 \le n < N-1$. Then the parameterization of $\mathcal{A}_n$ by $s \in (n, n+1)$ can be extended as a parameterization of $\overline{\mathcal{A}}_n$ by $s \in [n, n+1]$ with $(\Lambda_n(m), \kappa_n(m)) = (\lambda_m, x_m)$ when $m = n$ and $m = n+1$. Since $n < N - 1$, Definition 9.1.4 implies that $\overline{\mathcal{A}}_{n+1}$ can be parameterized, in a neighbourhood of (λ_n, x_n) by $s \in [n+1, n+1+\epsilon)$, for some $\epsilon > 0$, and so, in the first two cases above,

$$\begin{aligned}
(\Lambda(n), \kappa(n)) &= (\lambda_n, x_n), \ \ n = 0, 1, \cdots, N-1, \\
(\Lambda(s), \kappa(s)) &= (\Lambda_n(s), \kappa_n(s)) \text{ for } s \in (n, n+1),
\end{aligned}$$

defines a continuous parameterization

$$\{(\Lambda(s), \kappa(s)) : 0 \le s < N\} \tag{9.3}$$

of

$$\mathfrak{R} = \cup_{0 \leq n < N} \left(\mathcal{A}_n \cup \{(\lambda_n, x_n)\} \right) \subset \overline{\mathfrak{N}}.$$

Suppose that neither

$$\lim_{s \to N} \|(\Lambda(s), \kappa(s))\| = \infty \text{ nor } \lim_{s \to N} \text{dist}\big((\Lambda(s), \kappa(s)), \partial U\big) = 0 \qquad (9.4)$$

is true. Then there exists a sequence $t_k \to N$ with $\{(\Lambda(t_k), \kappa(t_k))\}$ both bounded in $\mathbb{R} \times X$ and bounded away from the boundary of U. Hence, from the compactness hypothesis of Theorem 9.1.1, $\{(\Lambda(t_k), \kappa(t_k))\}$ is relatively compact and, without loss of generality, we may suppose that it converges to $(\lambda^*, x^*) \in \mathcal{S}$, say.

If $N = \infty$ then every neighbourhood of (λ^*, x^*) intersects infinitely many distinct distinguished arcs, and this contradicts the fact that in a neighbourhood of (λ^*, x^*) the solution set is an $\mathbb{R}$-analytic variety. Now suppose that $N < \infty$ and $\overline{\mathcal{A}}_{N-1}$ is not compact. Since $t_k \to N$, $(\lambda^*, x^*) \in \partial \mathcal{A}_{N-1} \setminus \{(\lambda_{N-1}, x_{N-1})\}$. Using Lyapunov-Schmidt reduction we find that $\mathcal{A}_{N-1}$ in a neighbourhood of (λ^*, x^*) corresponds to a manifold of 1-regular points from a real-analytic one-dimensional branch of an analytic variety in a neighbourhood of $(\lambda^*, 0)$ in $\mathbb{R}^{q+1}$. By Corollary 7.5.3, as in the discussion preceding Definition 9.1.4, this contradicts the maximality of the route under consideration. Hence one of the alternatives in (9.4) occurs.

It is now straightforward to map $[0, N)$ to $[0, \infty)$ to obtain a parameterization of $\mathfrak{R}$ satisfying parts (a)–(d), and either (e)(i) or (e)(ii), in these two cases.

Next we consider the third case, when $N < \infty$ and $\overline{\mathcal{A}}_{N-1}$ is compact. Let (λ_{N-1}, x_{N-1}) and (λ_N, x_N) be the end points of $\overline{\mathcal{A}}_{N-1}$. The unique continuation of $\mathcal{A}_{N-1}$, as a distinguished arc $\mathcal{A}_N$ distinct from $\mathcal{A}_{N-1}$ with an end point in common with $\mathcal{A}_{N-1}$ at (λ_N, x_N), is ensured by Corollary 7.5.3 on the structure of one-dimensional varieties at a singular point. Since our route is maximal it follows from Definition 9.1.4 (d) that $\mathcal{A}_N = \mathcal{A}_m$ for some $m \in \{0, \cdots, N-2\}$.

Suppose that $m \in \{1, \cdots, N-3\}$. Since $\mathcal{A}_{m-1}$ and $\mathcal{A}_{m+1}$ are the only continuations of $\mathcal{A}_m$, and since $\mathcal{A}_{N-1}$ is a continuation of $\mathcal{A}_N$, it follows that $\mathcal{A}_{N-1} = \mathcal{A}_{m'}$ where $m' \in \{0, \cdots, N-2\}$. But this violates Definition 9.1.4 (d). Hence $\mathcal{A}_N = \mathcal{A}_0$ or $\mathcal{A}_N = \mathcal{A}_{N-2}$. In the latter case, it follows by induction that $\mathcal{A}_{2k} = \mathcal{A}_0$ and $\mathcal{A}_{2k+1} = \mathcal{A}_1$. Hence $N = 1$ and $\mathcal{A}_0 \cup \mathcal{A}_1$ forms a loop. On the other hand, when $\mathcal{A}_N = \mathcal{A}_0$, there are two possibilities, $(\lambda_N, x_N) = (\lambda_0, 0)$ or $(\lambda_N, x_N) = (\lambda_1, x_1)$. In the second of these cases it follows that $\mathcal{A}_{N-1} = \mathcal{A}_1$ which contradicts Definition 9.1.4 (d). The only remaining possibility is that $\mathcal{A}_N = \mathcal{A}_0$, $(\lambda_N, x_N) = (\lambda_0, 0)$ and $\mathfrak{R}$ is a loop in $\mathbb{R} \times X$ parameterized by (9.3). Once again we can parameterize $\mathfrak{R}$ by $s \in [0, \infty)$ so that parts (a)–(d) and (e)(iii) holds in this case.

Finally to prove (f). Suppose that $(\Lambda(s_1), \kappa(s_1)) = (\Lambda(s_2), \kappa(s_2))$, $s_1 \neq s_2$ and $\ker \partial_x F[(\Lambda(s_1), \kappa(s_1))] = \{0\}$. Then $(\Lambda(s_1), \kappa(s_1)) \in \mathcal{A}_{n_1}$ and $(\Lambda(s_2), \kappa(s_2)) \in \mathcal{A}_{n_2}$ for some $0 \leq n_1, n_2 < N$. From the implicit function theorem it follows that $\mathcal{A}_{n_1}$ and $\mathcal{A}_{n_2}$ coincide in a neighbourhood of the point where they intersect. Since $\mathcal{A}_{n_1} \cup \mathcal{A}_{n_2} \subset \mathfrak{N}$, the same argument gives that the maximal set where they coincide is $\mathcal{A}_{n_1} = \mathcal{A}_{n_2}$. Thus (e)(iii) occurs and $|s_1 - s_2|$ is an integer multiple of T.

This observation leads to the conclusion that $(\Lambda, \kappa) : [0, \infty) \to \mathcal{S}$ is locally injective which completes the proof of Theorem 9.1.1. $\qquad\qquad\qquad\qquad\square$

9.2 GLOBAL ANALYTIC BIFURCATION IN CONES

The following result, of which there are many variants, is important when positivity-invariant problems are under consideration. Such positivity, for example in problems of nonlinear elliptic partial differential equations, often follows from the maximum principle. The aim is to eliminate the possibility (e)(iii) that $\mathfrak{R}$ is a closed loop of solutions. If we can do this and $U = \mathbb{R} \times X$, then Theorem 9.1.1(e)(i) must occur and there exists a curve of solutions which becomes unbounded as the parameter tends to infinity (which is stronger than the statement that the curve is unbounded).

DEFINITION 9.2.1 *In a real Banach space X a closed set $\mathcal{K}$ is called a (non-convex) cone if $ax \in \mathcal{K}$ for all $a \geq 0$ and $x \in \mathcal{K}$.*

THEOREM 9.2.2 *In addition to the hypotheses of Theorem 9.1.1 suppose*

(a) *$\mathcal{K}$ is a cone in a real Banach space X.*

(b) *$\mathcal{R}^+ \subset \mathbb{R} \times \mathcal{K}$ (provided ϵ is chosen sufficiently small).*

(c) *If $\lambda \in \mathbb{R}$ and $\hat{\xi} \in \ker(\partial_x F[(\lambda, 0)]) \cap \mathcal{K}$, then $\hat{\xi} = \alpha \xi_0$ for $\alpha \geq 0$, and $\lambda = \lambda_0$. (In particular, $-\xi_0 \notin \mathcal{K}$.)*

(d) *Each point of $\mathfrak{R} \cap \mathcal{T} \cap (\mathbb{R} \times \mathcal{K})$ is an interior point of $\mathcal{T} \cap (\mathbb{R} \times \mathcal{K})$ in $\mathcal{S}$.*

Then $\kappa(s) \in \mathcal{K} \setminus \{0\}$ for all $s > 0$ and (e)(iii) in Theorem 9.1.1 does not occur.

Proof. Let $\bar{s} = \sup\{s > 0 : \kappa((0, s]) \subset \mathcal{K} \setminus \{0\}\}$ and suppose, for contradiction, that $\bar{s} < \infty$. Since $\mathcal{K}$ is closed, $\kappa(\bar{s}) \in \mathcal{K}$. Moreover $\kappa(\bar{s}) = 0$, since otherwise $\kappa(s) \in \mathcal{K}$ for some $s > \bar{s}$ by hypothesis (d) of the present theorem. Therefore $(\Lambda(\bar{s}), 0)$ is a bifurcation point on the line of trivial solutions of the equation $F(\lambda, x) = 0$. Since $(\Lambda(\bar{s}), \kappa(\bar{s})) \in \mathfrak{R}$, we can assume that, in a neighbourhood of $\bar{s}$, $\mathfrak{R}$ is parameterized $\mathbb{R}$-analytically by s. Let k denote the smallest natural number such that the kth derivative of κ at $s = \bar{s}$ is non-zero (k exists, by analyticity), so that

$$\kappa(s) = \kappa(s) - \kappa(\bar{s}) = \frac{d^k \kappa[\bar{s}]}{k!}(s - \bar{s})^k + O(|s - \bar{s}|^{k+1}).$$

Since, by definition of $\bar{s}$, $\kappa(s) \in \mathcal{K}$ for all s with $0 \leq s < \bar{s}$, we conclude that

$$(-1)^k d^k \kappa[\bar{s}] \in \mathcal{K} \setminus \{0\}.$$

Note that $\partial_{\lambda^m}^m F[(\lambda, 0)] = 0$ for all m. So differentiating the identity

$$F(\Lambda(s), \kappa(s)) = 0$$

k times at $s = \bar{s}$ leads to the conclusion that

$$(-1)^k d^k \kappa[\bar{s}] \in \ker\left(\partial_x F[(\Lambda(\bar{s}), \kappa(\bar{s}))]\right) \cap (\mathcal{K} \setminus \{0\}).$$

Now hypothesis (c) implies that $\Lambda(\bar{s}) = \lambda_0$ and $(-1)^k d^k \kappa[\bar{s}]$ is a positive multiple of ξ_0. Since λ_0 is a simple eigenvalue and $-\xi_0 \notin \mathcal{K}$, the bifurcating curve which lies in $\mathbb{R} \times \mathcal{K}$ and passes through the bifurcation point $(\lambda_0, 0)$ is uniquely determined as $\mathcal{R}^+$. Thus there is a segment of $\mathfrak{R}$, parameterized by $s < \bar{s}$ sufficiently close to $\bar{s}$, which is a subset of $\mathcal{R}^+$. Since $\mathcal{R}^+ \subset \mathcal{A}_0$, there exist sequences $\{s_k\}$, $\{t_k\}$ such that

$$(\Lambda(s_k), \kappa(s_k)) = (\Lambda(\bar{s} - t_k), \kappa(\bar{s} - t_k)), \quad s_k \searrow 0, \ t_k \searrow 0.$$

By Theorem 9.1.1 (f), $T > 0$ divides $\bar{s} - t_k - s_k$ for all k, which is false. Hence $\bar{s} = \infty$ and $\kappa(s) \in \mathcal{K} \setminus \{0\}$ for all $s > 0$. Since $(\lambda_0, 0) \in \mathfrak{R}$ we conclude that $\mathfrak{R}$ does not form a loop and the proof is complete. $\qquad\square$

9.3 BENDING AN ELASTIC ROD III

We now show how the theory of global bifurcation in cones can be applied to the boundary-value problem in §8.5:

$$\phi''(x) + \lambda \sin \phi(x) = 0 \text{ for } x \in [0, L], \quad \phi'(0) = \phi'(L) = 0, \qquad (9.5)$$

where L is fixed and $\lambda > 0$ is the parameter in the problem. As before let

$$\mathbb{F} = \mathbb{R}, \ X = \{\phi \in C^2[0, L] : \phi'(0) = \phi'(L) = 0\}, \ Y = C[0, L],$$

and define $F(\lambda, \phi) = \phi'' + \lambda \sin \phi$. Then $F : \mathbb{R} \times X \to Y$ is $\mathbb{R}$-analytic,

$$\partial_\phi F[(\lambda, 0)]\phi = 0$$

if and only if

$$\phi'' + \lambda\phi = 0 \in Y \text{ and } \phi'(0) = \phi'(L) = 0$$

and the bifurcation points form the set $\{\lambda_K = (K\pi/L)^2 : K \in \mathbb{N}\}$.

Here we focus on finding a global extension of the local bifurcation at the point $(\pi/L)^2$ corresponding to $K = 1$. In keeping with the notation of the last section let λ_0 denote $(\pi/L)^2$ and let $\xi_0(x) = \cos(\pi x/L)$, $x \in [0, L]$. Next we verify the hypotheses of Theorem 9.2.2. We have already seen that (G1) and (G3) hold. To check (G2) let $(\lambda, \psi) \in \mathbb{R} \times X$ be a solution of (9.5). Then

$$d_\phi F[(\lambda, \phi)](\psi) = \psi'' + \lambda\psi \cos \phi, \quad \psi \in X.$$

By the theory of ordinary differential equations, $\psi'' + \lambda\psi \cos\phi = 0$ has two linearly independent solutions at most one of which is in X. If there are no solutions in X the problem

$$\psi'' + \lambda\psi \cos\phi = f, \quad \psi \in X \tag{9.6}$$

has a solution ψ for every $f \in Y$. If, on the other hand, it has a solution $\hat{\psi} \in X$, then (9.6) has a solution if and only if

$$\int_0^L \hat{\psi}(x) f(x)\, dx = 0.$$

In both cases the range is closed, the codimension of the range and the dimension of kernel of $d_\phi F[(\lambda, \phi)]$ coincide.

This shows that in all cases $d_\phi F[(\lambda, \phi)]$ is a Fredholm operator of index zero and so (G2) holds.

Now let $\mathcal{K} \subset X$ be the cone defined by

$$\mathcal{K} = \{u \in X : u \text{ is odd about } L/2 \text{ and } u \geq 0 \text{ on } [0, L/2]\}.$$

We have seen that in this example hypothesis (a) of Theorem 9.2.2 holds, and (c) is obvious since the (unique up to normalization) eigenfunction corresponding the eigenvalue $(\pi K/L)^2$ is $\cos(K\pi x/L)$ and only when $K = 1$ is it in $\mathcal{K}$.

To see that (d) holds suppose that $(\lambda, \phi) \in \mathbb{R} \times (\mathcal{K} \setminus \{0\})$ satisfies (9.5). Then clearly $\lambda \neq 0$, $\sin\phi(0) \neq 0$ and $\sin\phi(L) \neq 0$. (If any one of them is zero then ϕ is a constant, by the uniqueness theorem for the initial-value problems for second order ordinary differential equations, and so ϕ is not odd about $L/2$.) Also any solution $(\hat{\lambda}, \hat{\phi})$ of (9.5) satisfies

$$\tfrac{1}{2}\hat{\phi}'(x)^2 + \hat{\lambda}\cos\hat{\phi}(0) - \hat{\lambda}\cos\hat{\phi}(x) \equiv 0 \text{ on } [0, L] \tag{9.7}$$

and, if $\hat{\lambda} \neq 0$, $\cos\hat{\phi}(0) = \cos\hat{\phi}(L)$.

Since $\lambda \neq 0$ and the derivative of cosine at $\phi(0)$ and at $\phi(L)$ is not zero, it follows that if $(\hat{\lambda}, \hat{\phi})$ is a solution of (9.5) which is sufficiently close to (λ, ϕ) then $\hat{\phi}(0) = -\hat{\phi}(L)$. Hence the functions $\hat{\phi}(x)$ and $-\hat{\phi}(L - x)$ solve the same initial value problem, and so are equal. This shows that $\hat{\phi}$ is odd about $L/2$.

Now to show that $\hat{\phi} \geq 0$ on $[0, L/2]$ suppose that there is a sequence (λ_k, ϕ_k) of solutions of (9.5) which converges to (λ, ϕ) in $\mathbb{R} \times X$ such that $\phi(x_k) < 0$, $x_k \in [0, L/2]$. Since $\phi_k(L/2) = 0$ and $\phi_k(0) > 0$ for k sufficiently large, we may assume that x_k is a minimizer of ϕ_k on $[0, L/2]$ and hence $\phi'_k(x_k) = 0$. In the limit as $k \to \infty$ we find that there exists $x \in [0, L/2]$ with $\phi(x) = \phi'(x) = 0$. By the uniqueness theorem for initial value problems this means that $\phi \equiv 0$, which is false. This contradiction establishes (d).

It remains to show (b), that $\mathcal{R}^+ \subset \mathbb{R} \times \mathcal{K}$. First we show that if $(\lambda, \phi) \in \mathcal{R}^+$, for $\epsilon > 0$ sufficiently small, then ϕ is odd about $L/2$. Recall from Theorem 8.4.1 that

$$\mathcal{R}^+ = \{(\Lambda(s), s(\xi_0 + \tau(s))) : s \in (0, \epsilon)\},$$

where $\Lambda(s) \to 1$ and $\tau(s) \to 0$ in X as $s \to 0$. To complete the proof that hypothesis (b) is satisfied recall that $\xi_0(x) = \cos(\pi x/L)$ and hence $\xi_0(0) = 1 = -\xi_0(L)$. So (9.7) gives

$$\cos\left(s(1+\tau(s)(0))\right) = \cos\left(s(-1+\tau(s)(L))\right) = \cos\left(s(1-\tau(s)(L))\right)$$

whence $s(1+\tau(s)(0)) = \pm s(1-\tau(s)(L))$ for $s > 0$ sufficiently small. It follows that the sign must be plus, and $\tau(s)(0) = -\tau(s)(L)$. Thus $s(\xi_0 + \tau(s))$ is odd about $L/2$ for $s > 0$ sufficiently small. Now $\kappa(s) = s(\xi_0 + \tau(s)) \geq 0$ on $[0, L/2]$ follows since $\kappa(L/2) = 0$, $\kappa(s)'(L/2) = s(-\pi/L + \tau(s)'(L/2))$ and $\tau(s) \to 0$ in X as $s \to 0$. Hence hypothesis (b) is satisfied.

Thus Theorem 9.2.2 gives the existence of a curve

$$\mathfrak{R} = \left\{(\Lambda(s), \kappa(s) : s \in [0, \infty)\right\}$$

with $(\Lambda(0), \kappa(0)) = ((\pi/L)^2, 0)$, $\kappa(s) \in \mathcal{K}$ for $s > 0$ and

$$\|(\Lambda(s), \kappa(s))\| \to \infty \text{ as } s \to \infty.$$

If now $(\lambda, \phi) = (\Lambda(s), \kappa(s)) \in \mathfrak{R}$ satisfies (9.5) it is obvious that $\lambda \neq 0$ and, by connectedness, $\lambda > 0$ for all $(\lambda, \phi) \in \mathfrak{R}$. Multiplying (9.5) by ξ_0 and integrating by parts gives

$$0 = \int_0^L \xi_0 \left(\phi'' + \sin\phi\right) dx = \int_0^L \phi\, \xi_0 \left(-\left(\frac{\pi}{L}\right)^2 + \frac{\lambda \sin\phi}{\phi}\right) dx$$

Since $\phi, \xi_0 \in \mathcal{K}$, the product $\phi\xi_0$ is non-negative and not identically zero. Since $\lambda > 0$ and $(\lambda \sin\phi)/\phi < \lambda$, it follows that $\lambda > (\pi/L)^2$ for all $(\lambda, \phi) \in \mathfrak{R}$, $\phi \neq 0$. Hence the global curve lies to the right of the bifurcation point. Since $\phi'(0) = 0 = \phi(L/2)$ for all solutions of (9.5), it is immediate that the set

$$\{(\lambda, \phi) \in \mathfrak{R} : \lambda \leq M\}$$

is bounded in $\mathbb{R} \times X$ for all finite M. Since $\mathfrak{R}$ is unbounded,

$$\left\{\lambda : (\lambda, \phi) = (\Lambda(s), \kappa(s)) : s > 0\right\} = ((\pi/L)^2, \infty).$$

Finally, if (λ, ϕ) is a solution of (9.5) ϕ can be extended as a smooth $2L$ periodic function on the real line. When this has been done, let

$$\mathfrak{R}_K = \left\{(K^2\lambda, \phi(Kx)) : (\lambda, \phi) \in \mathfrak{R}\right\}.$$

It is an easy matter to check that $\mathfrak{R}_K$ is a global branch of solutions bifurcating from $(KL/\pi)^2$, $K \in \mathbb{N}$.

Thus many qualitative features of the *global* bifurcation of solutions of (9.5), observed originally in the introduction, are a consequence of abstract considerations based on the theory of real analytic varieties and it is clear that the abstract method has much greater applicability. In the remaining chapters we give a substantial example to which the global theory makes a vital contribution.

9.4 NOTES ON SOURCES

The concept of global bifurcation has its origins in the work of Crandall and Rabinowitz [21] on ordinary differential equations, and was developed to include partial differential equations first by Rabinowitz [50] and then by Dancer [25] and Turner [64]. However, the approach for analytic operators is due to Dancer [26]. Some of the results given here appeared first in [13].

PART 4
Stokes Waves

Chapter Ten

Steady Periodic Water Waves

Now we embark on a study of global bifurcation in a problem from classical hydrodynamics. A steady periodic irrotational water wave of infinite depth, with a free surface under gravity and without surface tension, is called a Stokes wave. The first mathematical treatment of this free-boundary problem is the local existence theory due to Levi-Civita [42] and Nekrasov [47]. Although a breakthrough at the time, this is now recognised as an example of bifurcation from a simple eigenvalue.

The first global mathematical treatment is due to Krasov'skii [40], who obtained the existence of waves of all slopes from zero up to, but not including, $\pi/6$. However he did not show that they formed a connected set. That contribution was due to Keady and Norbury [36], who used the global topological bifurcation theory of Rabinowitz [50], as adapted for operator equations in cones by Dancer [25], to obtain the existence of a global connected set of Stokes waves.

Here we take matters further and show that there is a global curve of solutions which has a local analytic re-parameterization at every point and which connects the bifurcation point representing zero flow (no wave) to a Stokes wave of extreme form [57]. The present account is based on a formulation of the Stokes-wave problem as a pseudo-differential operator equation due originally to Babenko [8] (see [63] for further background) and draws upon theory developed in [4, 11, 12, 13, 44]; see also [14, 53, 62, 63]. We begin with a description of the physical problem and a derivation of the equation that is the focus of our study.

10.1 EULER EQUATIONS

The incompressible Euler equations for a velocity field $\vec{U}$ and a pressure field P in a force field $\vec{F}$ are

$$\vec{U}_t + (\vec{U} \cdot \nabla)\vec{U} = \nabla P + \vec{F}, \quad \nabla \cdot \vec{U} = 0.$$

In the next two sections we derive equations for steady irrotational water waves of infinite depth, with gravity but without surface tension.

Steady Euler Equations

We begin with a standard elliptic boundary-value problem on an unbounded domain.

DEFINITION 10.1.1   *A 2π-periodic function $u : \mathbb{R} \to \mathbb{R}$ is said to be Hölder continuous with exponent $\alpha \in (0,1)$, written $u \in C^\alpha$, if*

$$\|u\|_{C^\alpha} = \sup_{x \in [-\pi,\pi]} |u(x)| + \sup_{x,y \in [-\pi,\pi]} \frac{|u(x) - u(y)|}{|x-y|^\alpha} < \infty.$$

When u is k-times continuously differentiable on $(-\pi,\pi)$ and the kth derivative has continuous extension to $[-\pi,\pi]$ which is in C^α, we say that $u \in C^{k,\alpha}$. Note that a function u which is Lipschitz continuous on $[-\pi,\pi]$ need not be in C^1, the space of continuously differentiable functions on $[-\pi,\pi]$; instead we write $u \in C^{0,1}$ when u is Lipschitz continuous. In this notation, C^k, $k \in \mathbb{N}_0$, denotes the space of 2π-periodic, k-times-continuously differentiable functions on $\mathbb{R}$. This notation has an obvious extension to functions u defined on subsets U of $\mathbb{R}^m$, $C^{k,\alpha}(U)$, $C^{k,\alpha}(\overline{U})$ etc.

Note, for future reference that, as a consequence of the Ascoli-Arzelà theorem 2.7.1, $C^{k,\alpha}$, $\alpha \in (0,1)$, is compactly embedded in C^k, $k \in \mathbb{N}_0$.

Suppose w is a real-valued, 2π-periodic, even $C^{2,\alpha}$ function of a real variable and that a curve $\mathcal{S}$ and a set Ω are defined by

$$\mathcal{S} := \{(x, w(x)) : x \in \mathbb{R}\} \quad \text{and} \quad \Omega = \{(x,y) : y < w(x)\}.$$

For a real parameter c consider the boundary-value problem

$$\Delta \hat{\psi} = 0 \quad \text{in } \Omega, \quad \hat{\psi} = 0 \quad \text{on } \mathcal{S};$$

$$\hat{\psi} \ \text{is } 2\pi\text{-periodic and even in } x;$$

$$\nabla \hat{\psi}(x,y) - (0,c) \to 0 \quad \text{as } y \to -\infty.$$

For any $c > 0$ this problem has a unique solution which is real-analytic in Ω and all its derivatives and second derivatives have continuous extensions from Ω onto the boundary of Ω. To see this, a solution may be obtained by minimizing the functional

$$\int_{\Omega \cap ((-\pi,\pi) \times \mathbb{R})} |\nabla \hat{\psi}(x,y) - (0,c)|^2 \, dy dx$$

over the set of all functions in $W^{1,2}_{loc}(\Omega)$ which are 2π-periodic and even in x with zero trace on $\mathcal{S}$. Then the Phragèmen-Lindelöf principle (the maximum principle for harmonic functions on unbounded domains) gives that the solution is unique. The regularity of the solution to the boundary-value problem obtained in this way is ensured by standard theory [33, §6.4] which yields that $\hat{\psi} \in C^{2,\alpha}(\overline{\Omega})$.

Since $\hat{\psi} = 0$ on $\mathcal{S}$ and $\hat{\psi} \to -\infty$ as $y \to -\infty$, the harmonic function $\hat{\psi}$ is negative everywhere on Ω. It therefore attains its maximum on $\overline{\Omega}$ at every point of $\mathcal{S}$. Hence, by the boundary-point lemma,

$$\frac{\partial \hat{\psi}}{\partial y} > 0 \text{ on } \mathcal{S}.$$

Now note that the function $\partial\hat\psi/\partial y$ is a harmonic function on Ω which tends to $c > 0$ as $y \to -\infty$. The maximum principle therefore implies that $\partial\hat\psi/\partial y > 0$ everywhere on $\overline\Omega$. From the evenness and periodicity of $\hat\psi(\cdot, y)$ we infer that

$$\frac{\partial\hat\psi}{\partial x} = 0 \text{ at } x = 0, \pm\pi.$$

It follows from the implicit function theorem that, for any $\alpha < 0$, the set $\{(x, y) \in \Omega : \hat\psi(x, y) = \alpha\}$ is the graph of a smooth (in fact real-analytic) function Y_α which gives y as a function of x. Define a velocity field

$$\vec{U}(x, y, t) = (u(x, y, t), v(x, y, t)) = (\hat\psi_y, -\hat\psi_x).$$

Note that $\vec{U}$ is independent of time t and that div $\vec{U} = \nabla \cdot \vec{U} = 0$. In other words, $\vec{U}$ is a stationary and solenoidal. Let $\vec{F}_g = (0, -g) = \nabla(-gy)$ be the (constant) gravitational force field acting vertically downwards and define a scalar pressure field in Ω by

$$P(x, y) = \tfrac{1}{2}|\nabla\hat\psi(x, y)|^2 + gy.$$

Then $\vec{U}$ satisfies the Euler equations for a two-dimensional incompressible flow under gravity:

$$\vec{U}_t + (\vec{U} \cdot \nabla)\vec{U} = \nabla P + \vec{F}_g, \quad \nabla \cdot \vec{U} = 0.$$

Since curl $\vec{U} = \nabla \wedge \vec{U} = 0$, it is also irrotational. From the fact that $\hat\psi = 0$ on $\mathcal{S}$, and from the behaviour of $\nabla\hat\psi$ at infinite depth, we have that

$$0 = \nabla\hat\psi \cdot (1, w_x) = \hat\psi_x + \hat\psi_y w_x \text{ on } \mathcal{S},$$
$$\vec{U} \to (c, 0) \text{ as } y \to -\infty.$$

Since $\vec{U}$ is everywhere perpendicular to $\nabla\hat\psi$, it is tangent to the curve $\mathcal{S}$. Now define time-dependent domains Ω_t with boundaries $\mathcal{S}_t$, a velocity field $\vec{V}$ and a pressure $\mathcal{P}$ as follows:

$$\Omega_t = \{(x, y) : (x + ct, y) \in \Omega\}, \quad \eta(x, t) = w(x + ct),$$
$$\mathcal{S}_t = \{(x, y) : (x + ct, y) \in \partial\Omega\} = \{(x, y) \in \mathbb{R}^2 : y = \eta(x, t)\},$$
$$\vec{V}(x, y, t) = \vec{U}(x + ct, y, t) - (c, 0), \quad \mathcal{P}(x, y, t) = P(x + ct, y).$$

Since

$$\{\vec{V}_t + (\vec{V}.\nabla_{(x,y)})\vec{V}\}\big|_{(x,y,t)} = [c\vec{U}_x + (\vec{U} \cdot \nabla_{(x,y)}))\vec{U} - c\vec{U}_x]\big|_{(x+ct,y)}$$
$$= \{\nabla P + \vec{F}_g\}\big|_{(x+ct,y)} = \{\nabla_{(x,y)}\mathcal{P} + \vec{F}_g\}_{(x,y,t)},$$

$\vec{V}$, $\mathcal{P}$ define a steady solution of the Euler equations. Remember that $\mathcal{S}$, and hence $\mathcal{S}_t$, was specified at the outset and so we should think of $\mathcal{S}_t$ as a rigid boundary at time t which is moving horizontally from right to left with constant velocity c on the surface of an infinitely wide, infinitely deep, ocean under gravity. The flow it generates is described by $\vec{V}$ and $\mathcal{P}$ which satisfy the Euler equations, and $\vec{V}(x, y, t) \to 0$ as $y \to -\infty$.

Steady Symmetric Water Waves

If $\mathcal{S}$ and c are such that P is constant on $\mathcal{S}$, the rigid moving boundary is not needed to maintain the motion since the pressure $\mathcal{P}$ at the surface $\mathcal{S}_t$ is in equilibrium with constant atmospheric pressure.

Therefore steady symmetric periodic water waves are described by solutions of a boundary-value problem in which the domain Ω (described by a 2π-periodic even $C^{2,\alpha}$ function w), the function $\hat{\psi}$ and the parameter $c > 0$ are the unknowns:

$$\Delta\hat{\psi} = 0 \text{ in } \Omega; \tag{10.1a}$$

$$\hat{\psi}(x, w(x)) = 0 \text{ for all } x \in \mathbb{R}; \tag{10.1b}$$

$$\nabla\hat{\psi}(x, y) \to (0, c) \text{ as } y \to -\infty; \tag{10.1c}$$

$$\hat{\psi}(-x, y) = \hat{\psi}(x, y) = \hat{\psi}(x + 2\pi, y), \quad (x, y) \in \Omega; \tag{10.1d}$$

$$\tfrac{1}{2}|\nabla\hat{\psi}(x, w(x))|^2 + gw(x) \equiv \text{constant}, \ x \in \mathbb{R}. \tag{10.1e}$$

We have seen equations (10.1a)-(10.1d) already; the new ingredient here is (10.1e), which expresses the requirement that the pressure on the free surface is constant.

Dimensionless Variables

Let $\hat{\psi} = c\psi$. Then

$$\Delta\psi = 0 \text{ in } \Omega; \tag{10.2a}$$

$$\psi(x, w(x)) = 0 \text{ for all } x \in \mathbb{R}; \tag{10.2b}$$

$$\nabla\psi(x, y) \to (0, 1) \text{ as } y \to -\infty; \tag{10.2c}$$

$$\psi(-x, y) = \psi(x, y) = \psi(x + 2\pi, y), \quad (x, y) \in \Omega; \tag{10.2d}$$

$$\tfrac{1}{2}|\nabla\psi(x, w(x))|^2 + \lambda w(x) \equiv \tfrac{1}{2}, \ x \in \mathbb{R}, \ \lambda = g/c^2. \tag{10.2e}$$

There is no loss of generality in taking the constant on the right side of (10.2e) to be a half, since this can always be achieved by relocating the origin in the y direction, an operation which has no effect on the other equations. The parameter λ is a dimensionless parameter in the water-wave problem, the square of the Froude number.

Slight Generalization

For future reference consider briefly the more general problem in which the surface S, which is 2π-periodic and even in x, is given in parametric form by

$$S = \{(X(t), Y(t)) : t \in \mathbb{R}\},$$

where $t \mapsto (X(t), (Y(t))$ is a globally injective, $C^{2,\alpha}$ function, and (10.2b) and (10.2e) are satisfied on $\mathcal{S}$:

$$\psi(X(t), Y(t)) = 0 \text{ and } \tfrac{1}{2}|\nabla\psi(X(t), Y(t)))|^2 + \lambda Y(t) \equiv \tfrac{1}{2}, \ t \in \mathbb{R}.$$

LEMMA 10.1.2   *Suppose (X, Y) is a $C^{2,\alpha}$-function of t with $X'(t)^2 + Y'(t)^2 > 0$ and $X'(t) \geq 0$ on $\mathbb{R}$. Then $X'(t) > 0$ on $\mathbb{R}$.*

Proof. Since $(X(t), Y(t))$ is a $C^{2,\alpha}$ function of t, we know [33, §6.4] that ψ and all its first and second derivatives are continuous on $\overline{\Omega}$. Suppose that $X'(t_0) = 0$. Then the first boundary condition gives $\psi_y(X(t_0), Y(t_0)) = 0$ and

$$Y'(t_0)^2 \psi_{xx}(X(t_0), Y(t_0)) = \psi_x(X(t_0), Y(t_0)) X''(t_0). \tag{10.3}$$

Now $P = \tfrac{1}{2}|\nabla\psi|^2 + \lambda y$ is a sub-harmonic function on Ω which is constant on $\mathcal{S}$ and tends to $-\infty$ as $y \to -\infty$. By the Hopf boundary-point lemma its outward normal derivative on $\mathcal{S}$ is everywhere strictly positive. Since $\psi < 0$ on Ω, the outward normal is parallel to $\nabla\psi$ on the boundary and so

$$0 < \nabla\psi(X(t_0, Y(t_0)) \cdot \nabla P(X(t_0), Y(t_0))$$
$$= \psi_x(X(t_0), Y(t_0)) P_x(X(t_0), Y(t_0))$$
$$= \psi_x((X(t_0), Y(t_0))^2 \psi_{xx}(X(t_0), Y(t_0))$$

With (10.3), this gives that $X''(t_0) \neq 0$. Since $X'(t) \geq 0$ and $X'(t_0) = 0$ it follows that $X''(t_0) = 0$, which is a contradiction and the proof is complete.   □

Trivial Solution

For all $\lambda > 0$ the system (10.2) has a solution

$$w \equiv 0, \ \ \psi(x, y) = y, \ \ \mathcal{S} = \{(x, 0) : x \in \mathbb{R}\}, \ \ \Omega = \mathbb{R} \times (-\infty, 0),$$

which corresponds to the constant solution $\vec{U} = (c, 0)$ of the Euler equations and uniform parallel flow in a horizontal direction. This is a trivial solution.

10.2 ONE-DIMENSIONAL FORMULATION

To tackle the existence questions for non-trivial solutions using bifurcation theory we first need to re-formulate (10.2) as a nonlinear equation in the form $F(\lambda, w) = 0$, where $w : \mathbb{R} \to \mathbb{R}$ is 2π-periodic.

Let $-\phi$ be the harmonic conjugate of ψ in (10.2) so that $\phi + i\psi$ is holomorphic in Ω. Both ψ and ϕ are in $C^{2,\alpha}(\overline{\Omega})$. Since ψ is even, $\psi_x(x, y) = 0$ for all $(x, y) \in \Omega$ with $x \in \{0, \pm\pi\}$ and the Cauchy-Riemann equations ensure that we may normalize ϕ so that ϕ is odd and $\phi(\pm\pi, y)$ is independent of y. Since

$$\phi(\pi, y) - \phi(-\pi, y) = \int_{-\pi}^{\pi} \phi_x(x, y)dx = \int_{-\pi}^{\pi} \psi_y(x, y)dx \to 2\pi$$

as $y \to -\infty$, $\phi(\pm\pi, y) = \pm\pi$ for all $(\pm\pi, y) \in \Omega$ and $\phi(0, y) = 0$ for all $(0, y) \in \Omega$. Therefore, since $\psi_y \neq 0$ on $\overline{\Omega}$, a symmetric solution of the steady water-wave problem gives rise to a conformal bijection $\phi + i\psi$ from Ω onto $R := \mathbb{R} \times (-\infty, 0)$ and from $\Omega_\pi := \Omega \cap \{(-\pi, \pi) \times \mathbb{R}\}$ onto $R_\pi := (-\pi, \pi) \times (-\infty, 0)$. Let $S_\pi = \overline{\Omega}_\pi \cap S$. When composed with an exponential bijection this gives a conformal bijection ζ, defined on Ω_π by

$$\zeta(x + iy) = \exp\left(-i(\phi(x, y) + i\psi(x, y))\right),$$

which maps Ω_π onto $\mathcal{D} \setminus \{\zeta \in \mathbb{R} : \zeta \leq 0\}$, the open unit disc cut along the non-positive real axis (the point at infinity in Ω_π is mapped to the origin). Let Z denote its inverse, from $\mathcal{D} \setminus \{\zeta \in \mathbb{R} : \zeta \leq 0\}$ into the complex z-plane, $z = x + iy$. From the boundary conditions satisfied by ϕ and ψ, and from the behaviour of $\phi + i\psi$ as $y \to -\infty$, it follows that the function

$$Z(\zeta) - i \log \zeta, \quad \zeta \in \mathcal{D} \setminus \{\zeta \in \mathbb{R} : \zeta \leq 0\},$$

can be extended to the non-positive real axis to yield a complex-analytic function on $\mathcal{D}$. (Here log is the usual branch of logarithm which is real on the positive real axis.)

Therefore Z can be written in the form

$$Z(\zeta) = i\left(\log \zeta + \sum_{k=0}^{\infty} a_k \zeta^k\right) = i\left(\log \zeta + f(\zeta)\right), \tag{10.4}$$

say. The evenness of ψ with respect to x means that f is real when ζ is real and therefore all the coefficients a_k are real. Let points in the complex ζ-plane be identified by polar coordinates $\zeta = re^{it}$ (t no longer denotes time.) Then

$$Z(\zeta) = i \log r - t + \sum_{k=0}^{\infty} a_k r^k \left(i \cos kt - \sin kt\right). \tag{10.5}$$

It is clear that Ω_π, the image of Z on $\mathcal{D}$, is bounded by the vertical lines $x = \pm\pi$ and the curve S_π given parametrically by

$$x = -t - \sum_{k=1}^{\infty} a_k \sin kt, \quad y = \sum_{k=0}^{\infty} a_k \cos kt, \quad t \in [-\pi, \pi]. \tag{10.6}$$

Let us start again and suppose that f is a given holomorphic function on $\overline{\mathcal{D}}$ which is real on the real axis. Let Z be given in terms of f by (10.4). Suppose that Z is a bijection onto its range Ω_π and define $\Phi + i\Psi$ on Ω_π by

$$\Phi(Z(\zeta)) + i\Psi(Z(\zeta)) = i \log \zeta = i \log r - t \tag{10.7}$$

so that $\Psi \equiv 0$ on S_π. We now ask if this defines a solution of (10.2).

Since $\Im Z \to -\infty$ corresponds to $r \to 0$, we find from (10.4) and (10.7) that

$$\nabla\Phi(x + iy) \to (1, 0)$$

as $y \to -\infty$. It is therefore automatic that when Ψ is defined on Ω_π using (10.7) and the mapping f in (10.4), Ψ is harmonic and even with respect to x on Ω_π, $\Psi \equiv 0$ on $\mathcal{S}_\pi$ and $\nabla \Psi \to (0,1)$ as $y \to -\infty$. For Ψ to give a solution of the water-wave problem the only question remaining is whether

$$\tfrac{1}{2}|\nabla \Psi|^2 + \lambda y \equiv \tfrac{1}{2} \quad \text{on} \quad \mathcal{S}_\pi? \tag{10.8}$$

This requirement can be written as a further condition to be satisfied by f in (10.4) or, equivalently, by the coefficients $\{a_k\}$ in (10.5).

Since $(\Phi + i\Psi)(Z(\zeta)) = i \log \zeta$,

$$\frac{d}{dz}(\Phi + i\Psi)(Z(\zeta))\frac{dZ}{d\zeta} = \frac{i}{\zeta}$$

and hence

$$|\nabla \Psi(Z(\zeta))|^2 = |(\Phi + i\Psi)'(Z(\zeta))|^2 = \frac{1}{|\zeta|^2|Z'(\zeta)|^2}.$$

In particular, when $|\zeta| = 1$, so that $Z(\zeta) \in \mathcal{S}_\pi$ and $\zeta = e^{it}$, we have

$$|\nabla \Psi(Z(\zeta))|^2 = \frac{1}{|Z'(\zeta)|^2},$$

where

$$Z'(\zeta) = \frac{i}{\zeta} + i\sum_{k=1}^{\infty} ka_k\zeta^{k-1} = \frac{i}{\zeta}\left(1 + \sum_{k=1}^{\infty} ka_k\zeta^k\right).$$

Therefore (10.8) implies that at the point $\zeta = e^{it}$ on the unit circle,

$$|\nabla \Psi(Z(\zeta))|^2 = \frac{1}{\left|1 + \sum_{k=1}^{\infty} ka_k\zeta^k\right|^2}$$

$$= \left\{(1 + \sum_{k=1}^{\infty} ka_k \cos kt)^2 + (\sum_{k=1}^{\infty} ka_k \sin kt)^2\right\}^{-1}.$$

To write this in a convenient notation, define an operator $\mathcal{C}$ on $L_2[-\pi, \pi]$ by

$$\mathcal{C}(1) = 0, \quad \mathcal{C}(\sin kt) = -\cos kt, \quad \mathcal{C}(\cos kt) = \sin kt, \quad k \geq 1,$$

extended by linearity and continuity. Note from the Riesz-Fischer theorem that, as an operator on $L_2[-\pi, \pi]$, $\|\mathcal{C}\| \leq 1$ and that $\mathcal{C}^2 u = -u + [u]$, where $[u]$ denotes the mean of a function $u \in L_2[-\pi, \pi]$. The bounded linear operator $\mathcal{C}$ is the well known conjugation operator [69] from classical harmonic analysis.

Suppose that $w \in L_2[-\pi, \pi]$ is an even function with $w' = dw/dt \in L_2[-\pi, \pi]$ and Fourier series $\sum_{k=0}^{\infty} a_k \cos kt$. Then $w(-\pi) = w(\pi)$ and

$$1 + \mathcal{C}w'(t) = 1 + \sum_{k=1}^{\infty} ka_k \cos kt, \tag{10.9}$$

$$|\nabla\Psi(Z(e^{it}))|^2 = \frac{1}{(1 + \mathcal{C}w'(t))^2 + w'(t)^2}, \tag{10.10}$$

and, according to (10.6),

$$\mathcal{S}_\pi = \{(t + \mathcal{C}w(t), w(t)) : t \in [-\pi, \pi]\}. \tag{10.11}$$

Therefore the constant-pressure condition takes the form

$$(1 - 2\lambda w(t))\{w'(t)^2 + (1 + \mathcal{C}w'(t))^2\} = 1, \quad \text{for almost all } t \in [-\pi, \pi]. \tag{10.12}$$

Henceforth we will focus on (10.12). The argument which follows (10.6) shows that if $w \in C^{2,\alpha}$ is even and such that Z defined by (10.5) is globally injective and (10.12) holds then there exists a symmetric Stokes wave with profile given by

$$\{(t + \mathcal{C}w(t), w(t)) : t \in \mathbb{R}\}.$$

Conjugation Operator

To proceed we will need some standard operator theory. Everything in this section is proved in [9], [51] or [69].

Complex Analysis

In complex notation

$$\mathcal{C}e^{int} = -i \operatorname{sgn} n \, e^{int}, \quad n \in \mathbb{Z}, \tag{10.13}$$

with the convention that $\operatorname{sgn} 0 = 0$. Let u be a 2π-periodic, smooth, real-valued function with

$$u(t) = \sum_{n \in \mathbb{Z}} \hat{u}(n)e^{int}, \quad t \in [-\pi, \pi], \quad \hat{u}(-n) = \overline{\hat{u}(n)},$$

$$= \hat{u}(0) + \sum_{n \in \mathbb{N}} \hat{u}(n)e^{int} + \overline{\sum_{n \in \mathbb{N}} \hat{u}(n)e^{int}}$$

$$= \hat{u}(0) + \mathfrak{Re}\left(2 \sum_{n=1}^{\infty} \hat{u}(n)e^{int}\right).$$

Therefore u is the restriction $(u(t) = U(e^{it}))$ to the unit circle S^1 of the real part U of a complex holomorphic function f on the unit disc $\mathcal{D}$ in $\mathbb{C}$, where

$$f(z) = \hat{u}(0) + 2 \sum_{n \in \mathbb{N}} \hat{u}(n)z^n = U(z) + iV(z), \quad |z| \leq 1,$$

U and V are real and $V(0) = 0$. It follows that

$$V(e^{it}) = \Im f(e^{it}) = -i\Big\{ \sum_{n \in \mathbb{N}} \hat{u}(n)e^{int} - \overline{\sum_{n \in \mathbb{N}} \hat{u}(n)e^{int}} \Big\}$$

$$= -i\Big\{ \sum_{n \in \mathbb{N}} \hat{u}(n)e^{int} - \sum_{n \in \mathbb{N}} \hat{u}(-n)e^{-int} \Big\}$$

$$= -i\Big\{ \sum_{n \in \mathbb{Z}} \text{sgn}\, n\, \hat{u}(n)e^{int} \Big\} = \mathcal{C}u(t).$$

Thus the conjugation operation gives the restriction to S^1 of the imaginary part of a complex holomorphic function f on the closed unit disc when the restriction to the boundary of $\Re f$ is u and $\Im f(0) = 0$.

Further, by the Cauchy-Riemann equations in polar coordinates on the unit disc,

$$\mathcal{C}u'(t) = \frac{\partial U}{\partial r}\Big|_{e^{it}},$$

where

$$\Delta U = 0 \quad \text{on} \quad \mathcal{D}, \quad U(e^{it}) = u(t).$$

The above discussion is rigorous when the functions in question are smooth on the closed unit disc or on the unit circle. When u is square integrable, the theory is only a little more subtle.

Functional Analysis

For $u \in L_2[-\pi, \pi]$, $\mathcal{C}u$ (defined above in terms of Fourier coefficients) is given pointwise almost everywhere by the singular integral formula

$$\mathcal{C}u(t) = \frac{1}{2\pi} PV \int_{-\pi}^{\pi} \frac{u(s)ds}{\tan \frac{1}{2}(t - s)}, \tag{10.14}$$

where PV denotes a Cauchy principal value integral. Formula (10.14) defines $\mathcal{C}u$ pointwise almost everywhere for $u \in L_1[-\pi, \pi]$ and, although $\mathcal{C}$ maps neither $L_1[-\pi, \pi]$ nor $L_\infty[-\pi, \pi]$ into itself, we have the following.

THEOREM 10.2.1   (M. Riesz) $\mathcal{C} : L_p[-\pi, \pi] \to L_p[-\pi, \pi]$ *is a bounded linear operator for all* $p \in (1, \infty)$.

THEOREM 10.2.2   (Privalov) $\mathcal{C} : C^\alpha \to C^\alpha$ *is a bounded linear operator for all* $\alpha \in (0, 1)$.

THEOREM 10.2.3 *Suppose that* $u \in L_2[-\pi, \pi]$. *Then there exists a holomorphic function* f *defined on the unit disc such that, in* $L_2[-\pi, \pi]$ *and pointwise for almost all* $t \in [-\pi, \pi]$,

$$\lim_{r \to 1} f(re^{it}) = u(t) + i\mathcal{C}u(t).$$

Suppose that u, $w \in L_2[-\pi, \pi]$, α, $\beta \in \mathbb{R}$ and f, g are holomorphic in the unit disc with

$$\lim_{r \to 1} f(re^{it}) = u(t) + i(\alpha + \mathcal{C}u(t)), \quad \lim_{r \to 1} g(re^{it}) = w(t) + i(\beta + \mathcal{C}w(t)),$$

in $L_2[-\pi, \pi]$ and pointwise for almost all $t \in [-\pi, \pi]$. Then there exists $\gamma \in \mathbb{R}$ and a function $v \in L_1[-\pi, \pi]$ such that $\mathcal{C}v \in L_1[-\pi, \pi]$ and, in $L_1[-\pi, \pi]$ and pointwise for almost all $t \in [-\pi, \pi]$,

$$\lim_{r \to 1} f(re^{it})g(re^{it}) = v(t) + i(\gamma + \mathcal{C}v(t)).$$

In particular, if $fg|_{S^1}$ is imaginary, then $v = 0$ and $fg = i\gamma$ on $\mathcal{D}$ for some real constant γ.

Suppose that u is even and 2π-periodic. Let $\tilde{u}(t) = u(t + \pi)$. Then $\tilde{u}$ is even, 2π-periodic and

$$\mathcal{C}\tilde{u}(t) = (\mathcal{C}u)(t + \pi) = \widetilde{\mathcal{C}u}(t). \tag{10.15}$$

LEMMA 10.2.4 *Let $\mathcal{D}$ denote the unit disc in $\mathbb{C}$ and suppose that $f : \overline{\mathcal{D}} \to \mathbb{C}$ has the following properties:*
    $f \in C^{2,\alpha}(\overline{\mathcal{D}})$ is holomorphic in $\mathcal{D}$;
    $f|_{\partial \mathcal{D}}$ is injective so that $f(\partial \mathcal{D})$ is a simple closed Jordan curve Γ.
Let Ω be the interior of Γ, the bounded component of $\mathbb{C} \setminus \Gamma$. Then (a) $\Omega = f(\mathcal{D})$ and (b) $f : \overline{\mathcal{D}} \to \overline{\Omega}$ is a bijection.

Proof. Since f is non-constant on $\partial \mathcal{D}$, it is non-constant on $\mathcal{D}$. Since it is holomorphic on $\mathcal{D}$, f maps open sets to open sets. Therefore $\partial(f(\mathcal{D})) \subset f(\partial \mathcal{D}) = \Gamma$. If $f(\mathcal{D}) \not\subset \overline{\Omega}$, then there exists a path in $\mathbb{C} \setminus \overline{\Omega}$ jointing $x_0 \in f(\mathcal{D})$ to infinity. Therefore there is a point of $\partial(f(\mathcal{D}))$ which is not in $\overline{\Omega}$ and so not in Γ. This contradiction proves that $f(\mathcal{D}) \subset \overline{\Omega}$. Since $f(\mathcal{D})$ is open, $f(\mathcal{D}) \subset \Omega$. Now suppose that $f(\mathcal{D}) \neq \Omega$. Then there is a path in Ω joining a point of $f(\mathcal{D})$ to a point of Ω not in $f(\mathcal{D})$. Once again there is a point of $\partial(f(\mathcal{D}))$ not in Γ, which is a contradiction. This proves part (a).

To prove (b), note again that if $z_1 \in \partial \mathcal{D}$ and $z_2 \in \mathcal{D}$, then $f(z_1) \neq f(z_2)$, as in the argument for part (a). Suppose then that $z_1, z_2 \in \mathcal{D}$, $z_1 \neq z_2$ and $f(z_1) = f(z_2) = \omega \in \Omega$. Let $r \in (0, 1)$ be such that $|z_i| < r$, $i = 1, 2$. Then by [59, 3.4, page 115],

$$\frac{1}{2\pi i} \int_{\partial \mathcal{D}_r} \frac{f'(z)dz}{f(z) - \omega} \geq 2,$$

where $\mathcal{D}_r = \{z \in \mathcal{D} : |z| < r\}$. Since $|f(z) - \omega|$ is bounded away from 0 on $\partial \mathcal{D}$, and $f' \in L_1(\partial \mathcal{D})$, it follows (see, for example, [7, Thm. 6.6, page 100]) that

$$\frac{1}{2\pi i} \int_{\partial \mathcal{D}} \frac{f'(z)dz}{f(z) - \omega} \geq 2.$$

However,

$$\frac{1}{2\pi i} \int_{\partial D} \frac{f'(z)dz}{f(z) - \omega} = \frac{1}{2\pi i} \int_{\Gamma} \frac{d\zeta}{\zeta - \omega} = 1,$$

since $\omega \in \Omega$, the interior of Γ. This contradiction implies (b) and the lemma is proven.   □

10.3 MAIN EQUATION

To prove the existence of non-trivial solutions of equation (10.12) we replace it with another which is more convenient from many viewpoints. The relevant observation is the following.

DEFINITION 10.3.1 *Let Y denote the space of 2π-periodic absolutely continuous even functions $u : \mathbb{R} \to \mathbb{R}$ with $u' \in L_2[-\pi, \pi]$, endowed with the usual norm $\|u\|_Y^2 = \|u\|_{L_2[-\pi,\pi]}^2 + \|u'\|_{L_2[-\pi,\pi]}^2$. Let X and Z be the spaces of even and odd functions in $L_2[-\pi, \pi]$.*

THEOREM 10.3.2 *Suppose that $w \in Y$ is a solution of*

$$\mathcal{C}w' = \lambda\{w + w\mathcal{C}w' + \mathcal{C}(ww')\}. \tag{10.16}$$

(a) w satisfies (10.12).
(b) $1 - 2\lambda w > 0$ on $[-\pi, \pi]$ if and only if $w' \in L_3[-\pi, \pi]$.
(c) If $w \in C^{2,\alpha}$, $1 - 2\lambda w > 0$ and $1 + \mathcal{C}w' \geq 0$ on $[-\pi, \pi]$, then Z in (10.7) is injective and the argument following (10.6) means that the solution w of (10.16) gives rise to a symmetric Stokes wave with profile

$$\mathcal{S} = \{(t + \mathcal{C}w(t), w(t)) : t \in \mathbb{R}\} \text{ where } 1 + \mathcal{C}w' > 0 \text{ on } [-\pi, \pi].$$

Proof. (a) First re-write (10.16) in the form

$$(1 - 2\lambda w)(1 + \mathcal{C}w') + \mathcal{C}((1 - 2\lambda w)w') = 1 \tag{10.17}$$

and apply $\mathcal{C}$ to both sides to obtain

$$(1 - 2\lambda w)w' = \mathcal{C}((1 - 2\lambda w)(1 + \mathcal{C}w'))$$

so that

$$\mathcal{C}u = (1 - 2\lambda w)w' \text{ where } u = (1 - 2\lambda w)(1 + \mathcal{C}w').$$

Therefore, by Theorem 10.2.3, there exists a holomorphic function U on $\mathcal{D}$ with

$$U\big|_{S^1} = u + i\mathcal{C}u = (1 - 2\lambda w)\{1 + \mathcal{C}w' + iw'\}$$
$$= i(1 - 2\lambda w)\{w' - i(1 + \mathcal{C}w')\}.$$

By the same theorem there is a holomorphic function W on $\mathcal{D}$ such that $W|_{S^1} = w' + i(1 + Cw')$. Note also that the product UW has the property that

$$UW|_{S^1} = i(1 - 2\lambda w)\big(w'^2 + (1 + Cw')^2\big) \in i\mathbb{R}.$$

By the last part of Theorem 10.2.3, UW is constant on $\mathcal{D}$ and, by Cauchy's integral formula and (10.17),

$$U(0) = \frac{1}{2\pi} \int_{-\pi}^{\pi} (u + iCu)dt = \frac{1}{2\pi} \int_{-\pi}^{\pi} u\, dt$$

$$= \frac{1}{2\pi} \int_{-\pi}^{\pi} (1 - 2\lambda w)(1 + Cw')dt = 1.$$

Similarly $W(0) = i$, and therefore $UW \equiv i$ on $\mathcal{D}$. This identity restricted to S^1 gives

$$(1 - 2\lambda w)\{w'^2 + (1 + Cw')^2\} \equiv 1,$$

which is (10.12).

(b) Suppose that $w \in Y$ satisfies (10.16) and that $1 - 2\lambda w > 0$ everywhere. Since w is continuous there exists $\epsilon > 0$ such that $1 - 2\lambda w \geq \epsilon$. Since, by Theorem 10.3.2, w satisfies (10.12) it is immediate that $w' \in L_\infty[-\pi, \pi] \subset L_3[-\pi, \pi]$.

Now suppose that $w' \in L_3[-\pi, \pi]$. The M. Riesz theorem implies that $Cw' \in L_3[-\pi, \pi]$ also. Suppose, seeking a contradiction, that $1 - 2\lambda w(a) = 0$ for some $a \in [-\pi, \pi]$. Then, by Hölder's inequality,

$$|1 - 2\lambda w(t)| = 2\lambda \left| \int_a^t w'(s)ds \right| \leq 2\lambda \|w'\|_{L_3[-\pi,\pi]} |t - a|^{2/3}.$$

Hence, because w satisfies (10.12),

$$w'(t)^2 + (1 + Cw'(t))^2 \geq \frac{1}{2\lambda \|w'\|_{L_3[-\pi,\pi]} |t - a|^{2/3}}.$$

Since $|t - a|^{-1} \notin L_1[-\pi, \pi]$, $w'^2 + (1 + Cw')^2 \notin L_{3/2}[-\pi, \pi]$, which is a contradiction.

(c) From our hypotheses and (10.12), $t \mapsto (t + Cw(t), w(t))$, $t \in \mathbb{R}$ is an injective function with Hölder continuous second derivatives, by Privalov's theorem 10.2.2. It also follows from classical potential theory that ψ is C^2 on $\overline{\Omega}$. That Z is a global injection follows immediately from this observation and Lemma 10.2.4. Hence the argument following (10.6) gives the existence of a Stokes wave with symmetric free surface, in the notation of Lemma 10.1.2, given by

$$S = \{(X(t), Y(t)) : t \in \mathbb{R}\} = (t + Cw(t), w(t) : t \in \mathbb{R}\}.$$

Lemma 10.1.2 implies that $1 + Cw' > 0$ on $[-\pi, \pi]$ and the proof is complete. □

Some obvious features of (10.16) are the following.

(i) It is the Euler-Lagrange equation (Definition 3.4.2) of the functional

$$\mathcal{J}(w) = \int_{-\pi}^{\pi} \{w\mathcal{C}w' - \lambda w^2(1 + \mathcal{C}w')\} dt, \quad w \in Y.$$

(ii) It is a quadratic equation (with no higher order terms) the non-zero solutions of which give rise to exact solutions of the steady periodic water-wave problem (without approximation).

(iii) It has the trivial solution $w = 0$ for all λ. Linearizing with respect to w about the trivial solution yields the self-adjoint eigenvalue problem

$$\mathcal{C}w' = \lambda w, \quad w \in Y.$$

Since Y consists of even functions, the linearized problem has a complete set of eigenfunctions $\{\cos nt\}_{n \in \mathbb{N} \cup \{0\}}$, the corresponding set of eigenvalues being $\mathbb{N} \cup \{0\}$.

(iv) It can be re-written

$$(1 - 2\lambda w)\mathcal{C}w' = \lambda\{w - \mathcal{Q}(w)\}, \tag{10.18}$$

or as

$$\mathcal{C}\big((1 - 2\lambda w)w'\big) = \lambda\{w + \mathcal{Q}(w)\} \tag{10.19}$$

where $\mathcal{Q}(u) = u\mathcal{C}u' - \mathcal{C}(uu')$. (See §10.5 for properties of $\mathcal{Q}(u)$.)

(v) It can be re-written as $G(\lambda, w) = 0$, where $G : (0, \infty) \times Y \to X$ is defined by

$$G(\lambda, w) = \mathcal{C}w' - \lambda\{w + w\mathcal{C}w' + \mathcal{C}(ww')\}. \tag{10.20}$$

For $(\lambda, u) \in \mathcal{U}$ where

$$\mathcal{U} = \{(\lambda, w) \in (0, \infty) \times Y : 1 - 2\lambda w > 0\}, \tag{10.21}$$

$\partial_w G[(\lambda, w)] : Y \to X$ is Fredholm with index 0. (This is proved in §10.5.)

All these features are favourable; an awkward aspect of (10.16) is the involvement of the conjugation operator which is non-local and there is no obvious analogue of the maximum principle. To extract *a priori* bounds on solutions of (10.16) we introduce the classical equation for Stokes waves.

10.4 *A PRIORI* BOUNDS AND NEKRASOV'S EQUATION

Let w be an even $C^{2,\alpha}$ function such that (10.16), and hence (10.12), is satisfied. By Theorem 10.2.3 there exists a holomorphic function U on the unit disc $\mathcal{D}$ with $U(e^{it}) = 1 + Cw'(t) - iw'(t)$. Suppose in addition that $1 + Cw' > 0$ on $[-\pi, \pi]$. Then U is nowhere zero on $\mathcal{D}$ and its logarithm is well-defined. In polar coordinates let

$$\log U(re^{it}) = \log |U(re^{it})| + i\Theta(re^{it}) \text{ where } \Theta(e^{it}) = \vartheta(t).$$

By (10.12),

$$\log U(e^{it}) = \log |U(e^{it})| + i\vartheta(t) = -\tfrac{1}{2}\log(1 - 2\lambda w(t)) + i\vartheta(t),$$

$$\text{where } \vartheta(t) \in (-\tfrac{1}{2}\pi, \tfrac{1}{2}\pi) \text{ and } \tan\vartheta(t) = \frac{-w'(t)}{1 + Cw'(t)}, \ t \in [-\pi, \pi].$$

By Privalov's theorem 10.2.2, $\vartheta' \in C^{1,\alpha}$ since $w \in C^{2,\alpha}$. By the Cauchy-Riemann equations,

$$\frac{\partial\Theta}{\partial r}\bigg|_{e^{it}} = \frac{\partial}{\partial t}\left(\tfrac{1}{2}\log(1 - 2\lambda w(t))\right) = \frac{-\lambda w'(t)}{1 - 2\lambda w(t)}$$

$$= \lambda (1 - 2\lambda w(t))^{-3/2} \sin\vartheta(t).$$

On the other hand $\sin\vartheta(t) = -w'(t)\sqrt{1 - 2\lambda w(t)}$ gives

$$3\lambda\left\{\nu + \int_0^t \sin\vartheta(s)\,ds\right\} = (1 - 2\lambda w(t))^{3/2}. \tag{10.22}$$

The new parameter ν, which is related to (λ, w) in (10.16) by

$$\nu = (1 - 2\lambda w(0))^{3/2}/3\lambda,$$

is called Nekrasov's parameter,

$$\Theta(e^{it}) = \vartheta(t) \text{ and } \frac{\partial\Theta}{\partial r}\bigg|_{e^{it}} = \frac{\sin\vartheta(t)}{3\left\{\nu + \int_0^t \sin\vartheta(s)\,ds\right\}} \tag{10.23}$$

Since w is even, ϑ is odd. Note that the parameter λ does not appear in (10.23). Moreover the favourable properties (i)–(v) are lost. Nevertheless it leads to the following observation which will be useful later.

LEMMA 10.4.1 *Suppose w is a non-constant, even, $C^{2,\alpha}$ function which satisfies* (10.16) *with*

$$1 + Cw' \geq 0 \text{ and } 1 + Cw' + |w'| > 0 \text{ on } [-\pi, \pi], \ w' \leq 0 \text{ on } (0, \pi).$$

Then $1 + Cw' > 0$ on $[-\pi, \pi]$, $w' < 0$ on $(0, \pi)$, $w''(0) < 0 < w''(\pi)$ and $\vartheta'(0) > 0 > \vartheta'(\pi)$.

Proof. That $1 + Cw' > 0$ on $[-\pi, \pi]$ is immediate from Theorem 10.3.2 (c) and Lemma 10.1.2. Hence it suffices to prove that $0 \not\equiv \vartheta \geq 0$ on $(0, \pi)$ implies that $\vartheta > 0$ on $(0, \pi)$. Since Θ is continuous on $\overline{\mathcal{D}}$ and odd in t, it follows that Θ is non-negative on the boundary of the upper half disc $\mathcal{D}^+ = \{re^{it} : t \in [0, \pi], r \in [0, 1]\}$. If $\vartheta(t_0) = 0$, $t_0 \in (0, \pi)$, then, by the maximum principle, Θ has a minimum on $\overline{\mathcal{D}^+}$ at e^{it_0} and, by the Hopf boundary-point lemma, $\partial\Theta/\partial r(e^{it_0}) < 0$. However, by (10.23), this is a contradiction which shows that $w' < 0$ on $(0, \pi)$.

Since $\partial\Theta/\partial r = 0$ on the real axis in $\mathcal{D}$ and $(\partial\Theta/\partial r)(e^{it}) > 0$, $t \in (0, \pi)$, by (10.23), the maximum principle gives that the harmonic function $r\partial\Theta/\partial r$ is positive on $\mathcal{D}^+$ and takes its minimum on $\overline{\mathcal{D}^+}$ at every point of $\partial\mathcal{D}$ on the real axis. The Hopf boundary-point lemma gives that

$$\frac{\partial^2\Theta}{\partial t\partial r}(r) > 0 > \frac{\partial^2\Theta}{\partial t\partial r}(-r), \quad r \in (0, 1).$$

Since $(\partial\Theta/\partial t)(\pm r) \to 0$ as $r \to 0$ and $w'(0) = w'(\pi) = 0$,

$$\frac{-w''(0)}{1 + Cw'(0)} = \sec^2\Theta(1)\frac{\partial\Theta}{\partial t}(1) > 0$$

$$0 > \sec^2\Theta(-1)\frac{\partial\Theta}{\partial t}(-1) = \frac{-w''(\pi)}{1 + Cw'(\pi)}.$$

This completes the proof.   □

Therefore, when w is even and $1 + Cw' \geq 0$, ϑ is odd and $0 < \vartheta < \frac{1}{2}\pi$ on $(0, \pi)$. So, by elementary Fourier-series methods, the boundary-value problem (10.23) is equivalent to an integral equation for the odd function $\vartheta \in C^{1,\alpha}$. On $[0, \pi]$ it must satisfy

$$\vartheta(t) = \frac{1}{3}\int_0^\pi K(t, s)\frac{\sin\vartheta(s)}{\nu + \int_0^s \sin\vartheta(\tau)d\tau}ds, \tag{10.24a}$$

$$0 < \vartheta(t) < \tfrac{1}{2}\pi, \ t \in (0, \pi), \tag{10.24b}$$

where $\nu > 0$ and

$$K(t, s) = \frac{2}{\pi}\sum_{k=1}^\infty \frac{\sin kt \sin ks}{k} = \frac{1}{\pi}\log\left|\frac{\sin\frac{1}{2}(s+t)}{\sin\frac{1}{2}(s-t)}\right| > 0. \tag{10.25}$$

REMARK 10.4.2 The sine-series expression for K is immediate from the Fourier-series method used to derive (10.24) from (10.23). For later use we now derive the close-form formula (10.25) which gives the positivity of K. Recall (see, for example, [38, Ch. XII, §55, V]) that for $z \in \mathbb{C}$ with $|z| \leq 1$ and $z \neq 1$,

$$-\log(1 - z) = \sum_{k=1}^\infty \frac{z^k}{k},$$

where log here means the principal logarithm. In particular, for $x \in (0, 2\pi)$,

$$\sum_{k=1}^{\infty} \frac{\cos kx}{k} = -\log|2\sin\tfrac{1}{2}x| \text{ and } \sum_{k=1}^{\infty} \frac{\sin kx}{k} = \frac{1}{2}(\pi - x). \tag{10.26}$$

Hence, for $s, t \in [0, \pi]$, $s \neq t$,

$$\sum_{k=1}^{\infty} \frac{\sin kt \sin ks}{k} = \lim_{N\to\infty} \sum_{k=1}^{N} \frac{\cos k(t-s) - \cos k(t+s)}{2k}$$

$$= \frac{1}{2}\log\left|\frac{\sin\tfrac{1}{2}(t+s)}{\sin\tfrac{1}{2}(t-s)}\right| > 0,$$

which gives (10.25). Now by (10.26)

$$\int_0^\pi \frac{\sin kt}{k}\cot\tfrac{1}{2}t\,dt = -2\int_0^\pi \cos kt\log|2\sin\tfrac{1}{2}t|dt = \frac{\pi}{k}.$$

Therefore

$$\int_0^\pi K(s,t)\cot\tfrac{1}{2}t\,dt = 2\sum_{k=1}^{\infty} \frac{\sin ks}{k} = \pi - s, \quad s \in (0, \pi) \tag{10.27}$$

and (10.32) below follows from the symmetry of K. $\qquad\square$

Suppose that $\nu > 0$ and ϑ satisfy (10.24). Then $\vartheta(0) = \vartheta(\pi) = 0$ and either ϑ is identically zero or $\vartheta(x) > 0$ on $(0, \pi)$ because of the positivity of the kernel K. Moreover, multiplying the equation by $\sin x$ and integrating over $(0, \pi)$ using (10.25) yields that

$$\int_0^\pi \vartheta(x)\sin x\,dx = \frac{1}{3}\int_0^\pi \frac{\sin\vartheta(t)\sin t\,dt}{\nu + \int_0^t \sin\vartheta(w)dw}$$

$$< \frac{1}{3\nu}\int_0^\pi \vartheta(x)\sin x\,dx,$$

which implies $\quad 0 < \nu < 1/3$. $\tag{10.28}$

LEMMA 10.4.3 *Suppose that ϑ satisfies (10.24) on $[0, \pi]$. Then there exists $c > 0$ (independent of ϑ, ν and t) such that*

$$\nu + \int_0^t \sin\vartheta(\tau)d\tau \geq ct, \quad t \in [0, \pi].$$

Proof. Note that $x^{-1}\sin x$ is decreasing on $(0, \pi)$ and $x^{-\beta}\sin x$ is increasing on $(0, \pi/3)$ for $\beta = \pi/3\sqrt{3}$. Hence

$$\frac{2}{\pi}\vartheta(t) \leq \sin\vartheta(t) \leq \vartheta(t), \quad t \in (0, \tfrac{1}{2}\pi)$$

and, for s, $t \in (0, \pi/3)$,

$$\frac{\pi}{3\sqrt{3}} \log \left| \frac{s+t}{s-t} \right| \leq \log \left| \frac{\sin \frac{1}{2}(s+t)}{\sin \frac{1}{2}(s-t)} \right| \leq \log \left| \frac{s+t}{s-t} \right|.$$

Therefore, by (10.24), for $x \in (0, \pi/6)$

$$\int_x^{2x} \frac{\vartheta(t)\, dt}{t}$$

$$= \frac{1}{3\pi} \int_0^\pi \frac{\sin \vartheta(s)}{\nu + \int_0^s \sin \vartheta(\tau)d\tau} \left\{ \int_x^{2x} \frac{1}{t} \log \left| \frac{\sin \frac{1}{2}(s+t)}{\sin \frac{1}{2}(s-t)} \right| dt \right\} ds$$

$$\geq \frac{2}{9\pi\sqrt{3}} \int_x^{2x} \frac{\vartheta(s)}{\nu + \int_0^s \vartheta(\tau)d\tau} \left\{ \int_x^{2x} \frac{1}{t} \log \left| \frac{s+t}{s-t} \right| dt \right\} ds$$

$$= \frac{2}{9\pi\sqrt{3}} \int_x^{2x} \frac{\vartheta(s)}{\nu + \int_0^s \vartheta(\tau)d\tau} \left\{ \int_{x/s}^{2x/s} \frac{1}{u} \log \left| \frac{1+u}{1-u} \right| du \right\} ds$$

$$\geq \frac{\log 3}{18\pi\sqrt{3}} \int_x^{2x} \frac{\vartheta(s)}{\nu + \int_0^s \vartheta(\tau)d\tau}\, ds$$

$$= \frac{\log 3}{18\pi\sqrt{3}} \log \left\{ 1 + \frac{\int_x^{2x} \vartheta(\tau)d\tau}{\nu + \int_0^x \vartheta(\tau)d\tau} \right\},$$

since the interval $[x/s, 2x/s]$ has length at least a half when $s \in [x, 2x]$ and so

$$\int_{x/s}^{2x/s} \frac{1}{u} \log \left| \frac{1+u}{1-u} \right| du \geq \frac{1}{2} \min_{u \in [\frac{1}{2}, 2]} \left\{ \frac{1}{u} \log \left| \frac{1+u}{1-u} \right| \right\} = \frac{1}{4} \log 3.$$

Since $\vartheta < \frac{1}{2}\pi$,

$$\int_x^{2x} \frac{\vartheta(\theta)}{t}\, dt \leq \frac{1}{2}\pi \log 2,$$

and it follows from the above inequality that

$$1 + \frac{\int_x^{2x} \vartheta(\tau)d\tau}{\nu + \int_0^x \vartheta(\tau)d\tau}$$

is bounded, by M, say, independent of ϑ, ν and x. Since there exists $K > 0$ such that for all $m \in [0, M]$, $\log(1 + m) \geq Km$,

$$\int_x^{2x} \frac{\vartheta(\theta)}{t}\, dt \geq K \left\{ \frac{\int_x^{2x} \vartheta(\tau)d\tau}{\nu + \int_0^x \vartheta(\tau)d\tau} \right\} \geq K \left\{ \frac{x \int_x^{2x} \frac{\vartheta(\tau)}{\tau} d\tau}{\nu + \int_0^x \vartheta(\tau)d\tau} \right\},$$

where $K > 0$ changes at each step but is independent of ν, ϑ and x. This proves the result for $t \in [0, \pi/6]$ and the lemma follows. $\square$

THEOREM 10.4.4 *Let $\vartheta \in C^{1,\alpha}$ satisfy (10.24). Then*

$$0 > t\vartheta'(t) - \vartheta(t), \quad t \in (0, \pi),$$
$$\vartheta(t) < \pi/3, \quad t \in [0, \pi],$$
$$0 < k \le \lambda \le K$$
$$-K \le w \le K$$
$$\|w'\|_{L_p[-\pi, \pi]} \le K_p < \infty \text{ for } p < 3,$$

where k, K and K_p are constants independent of λ, w, ϑ, ν.

Proof. Note that $\vartheta > 0$ on $(0, \pi)$, $\vartheta'(0) > 0 > \vartheta'(\pi)$ and $\vartheta'(t) = \vartheta'(0) + N(t)$ where $|N(t)| \le \text{const } t^\alpha$. Hence for any $A > 1$, $t\vartheta'(t) - A\vartheta(t) < 0$ for t in a left neighbourhood of π and, for t in a right neighbourhood of 0,

$$t\vartheta'(t) - A\vartheta(t) = (1 - A)t\vartheta'(0) + tN(t) - A\int_0^t N(s)ds < -Ct$$

if $A > 1$ where $C > 0$ is a constant.

Suppose that the first inequality of the theorem is false. One possibility is that

$$t\vartheta'(t) - \vartheta(t) \le 0 \text{ on } [-\pi, \pi] \text{ and } t_0\vartheta'(t_0) - \vartheta(t_0) = 0, \quad t_0 \in (0, \pi).$$

If this is not the case then

$$1 < A = \inf\{a \ge 1 : t\vartheta'(t) - a\vartheta(t) \le 0, \ t \in (0, \pi)\}$$

and, from the behaviour of $t\vartheta'(t) - \vartheta(t)$ in a neighbourhood of 0 and π, there exists $t_0 \in (0, 1)$ with $t_0\vartheta'(t_0) - A\vartheta(t_0) = 0$. Hence, in all cases, there exists $A \ge 1$ and $t_0 \in (0, \pi)$ such that

$$t\vartheta'(t) - A\vartheta(t) \le 0 \text{ on } [-\pi, \pi] \text{ and } t_0\vartheta'(t_0) - A\vartheta(t_0) = 0.$$

As noted in the proof of Lemma 10.4.1, $\Theta > 0$ on $\mathcal{D}^+$, and hence

$$\left.\frac{\partial\Theta}{\partial t}\right|_{re^{i\pi}} \le 0, \quad 0 < r \le 1. \tag{10.29}$$

Let Θ_t denote the harmonic function on $\mathcal{D}$ which vanishes at the origin defined by

$$\Theta_t(r^{it}) = \sum_{n=1}^\infty n\, a_n r^n \cos nt \quad \text{when} \quad \Theta(re^{it}) = \sum_{n=1}^\infty a_n r^n \sin nt$$

and let $U : \overline{\mathcal{D}^+} \to \mathbb{R}$ be defined by

$$U(re^{it}) = t\Theta_t(re^{it}) - A\Theta(re^{it}).$$

From a bootstrap argument based on (10.24), it follows easily that ϑ is infinitely differentiable, and hence that Θ is infinitely differentiable on $\overline{\mathcal{D}^+}$. So U is smooth on $\mathcal{D}^+$ and a calculation reveals that U satisfies the equation

$$\Delta U - \frac{2}{t}U_t - \frac{A-1}{t^2}U = \frac{2A(A-1)\Theta}{t^2} \ge 0 \text{ on } \mathcal{D}^+. \tag{10.30}$$

Also

$$U(re^{i0}) = 0 \text{ and } U(re^{i\pi}) \leq 0, \ r \in (0,1], \ \text{by (10.29)} . \tag{10.31}$$

Now (10.30) is an elliptic equation (with coefficients that tend to infinity at certain points on the boundary of $\mathcal{D}^+$) which is satisfied by U on $\mathcal{D}^+$. By the maximum principle, applied on open balls interior to $\mathcal{D}^+$, the maximum of U on $\overline{\mathcal{D}^+}$ is attained on the boundary of $\mathcal{D}^+$. By (10.31) and the hypothesis of the theorem, U is negative in $\mathcal{D}^+$ and has a maximum on $\overline{\mathcal{D}^+}$ at e^{it_0}, $t_0 \in (0, \pi)$ where $U(e^{it_0}) = 0$.

The Hopf boundary-point lemma and (10.23) imply that

$$0 < \frac{\partial U}{\partial r}(e^{it_0}) = t_0 \frac{\partial \Theta_t}{\partial r}(e^{it_0}) - A \frac{\partial \Theta}{\partial r}(e^{it_0}) =$$

$$\frac{t_0}{3} \left(\frac{\vartheta'(t_0) \cos \vartheta(t_0)}{\nu + \int_0^{t_0} \sin \vartheta d\tau} - \frac{\sin^2 \vartheta(t_0)}{\left(\nu + \int_0^{t_0} \sin \vartheta d\tau\right)^2} \right) - \frac{A \sin \vartheta(t_0)}{3\left(\nu + \int_0^{t_0} \sin \vartheta d\tau\right)}$$

$$= \frac{A}{3} \left(\frac{\vartheta(t_0) \cos \vartheta(t_0) - \sin \vartheta(t_0)}{\nu + \int_0^{t_0} \sin \vartheta d\tau} \right) - \frac{t_0 \sin^2 \vartheta(t_0)}{3\left(\nu + \int_0^{t_0} \sin \vartheta d\tau\right)^2} < 0,$$

since $\vartheta \cos \vartheta - \sin \vartheta < 0$. This contradiction proves the first inequality and means that $\sin \vartheta(t)/t$ is decreasing on $(0, \pi]$. Therefore

$$\frac{\sin \vartheta(t)}{\nu + \int_0^t \sin \vartheta(\tau) \, d\tau} < \frac{\sin \vartheta(t)}{\int_0^t \tau \left(\sin \vartheta/\tau \right) d\tau}$$

$$< \frac{2}{t} < \frac{2}{\sin t} = \tan \tfrac{1}{2}t + \cot \tfrac{1}{2}t.$$

Since, by (10.27),

$$\int_0^\pi K(s,t) \tan \tfrac{1}{2}t \, dt = s \text{ and } \int_0^\pi K(s,t) \cot \tfrac{1}{2}t \, dt = \pi - s, \tag{10.32}$$

the bound on ϑ follows.

Since $\mathcal{C}w'$ has zero mean, the definition of ϑ and (10.22) give that

$$2\pi = \int_{-\pi}^\pi 1 + \mathcal{C}w' \, dt = \int_{-\pi}^\pi \frac{\sqrt{1 - 2\lambda w} \, (1 + \mathcal{C}w')}{\sqrt{1 - 2\lambda w}} dt$$

$$= \int_{-\pi}^\pi \frac{\cos \vartheta}{\sqrt{1 - 2\lambda w}} dt = \int_{-\pi}^\pi \frac{\cos \vartheta}{\left[3\lambda\{\nu + \int_0^t \sin \vartheta(s) \, ds\}\right]^{1/3}} dt.$$

Since $\nu < 1/3$ by (10.28), the bound on λ follows from Lemma 10.4.3 and the bound on ϑ. This calculation also gives that

$$(1 - 2\lambda w(0))^{3/2} = 3\lambda\nu < 1.$$

Hence $w(0) > 0$. On the other hand, an integration of (10.16) yields that

$$\int_{-\pi}^\pi w(t)dt = -\lambda \int_{-\pi}^\pi w(t)\mathcal{C}w'(t)dt < 0.$$

Hence there is a point in $(0, \pi)$ where w is zero.

The $L_p[-\pi, \pi]$-bound on w' follows from (10.12), (10.22) and Lemma 10.4.3. This in turn yields the bound on w and the proof is complete. $\qquad\square$

10.5 WEAK SOLUTIONS ARE CLASSICAL

For w to satisfy (10.16) or (10.12) it is sufficient for w to have a square-integrable derivative (or merely to be absolutely continuous). On the other hand, in §10.1 the profile $(X(t), Y(t))$, $t \in \mathbb{R}$, of a steady wave has $w \in C^{2,\alpha}$. Thus §10.2 contains a discussion of weak solutions (which we prefer because of the favourable properties (i)–(v) listed there) to the classical problem discussed in §10.1. We now show that provided $1 - 2\lambda w > 0$ every weak solution is a classical solution.

The Operator $\mathcal{Q}$

Formula (10.13) for the conjugation operator leads to the observation that $u \mapsto \mathcal{C}u'$ is an unbounded, densely defined, self-adjoint operator on $L_2[-\pi, \pi]$ given in terms of Fourier coefficients by $\hat{u}(k) \mapsto |k|\hat{u}(k)$, where $\hat{u}(k)$ denotes the kth Fourier coefficient of u, $k \in \mathbb{Z}$. The Cauchy principle value formula for $\mathcal{C}$ leads to the following observation which plays a central rôle in the Stokes-wave problem.

For a periodic function u with $u' \in L_2[-\pi, \pi]$ let $\mathcal{Q}(u)$ denote the function

$$\mathcal{Q}(u)(t) = u(t)\mathcal{C}u'(t) - \mathcal{C}(uu')(t).$$

LEMMA 10.5.1 *Suppose that u is 2π-periodic and $u' \in L_2[-\pi, \pi]$. Then for almost all $t \in \mathbb{R}$*

$$0 \le \mathcal{Q}(u)(t) = \frac{1}{8\pi} \int_{-\pi}^{\pi} \left\{ \frac{u(t) - u(s)}{\sin \frac{1}{2}(t - s)} \right\}^2 ds \le \frac{2}{\pi} \|u'\|^2_{L_2[-\pi,\pi]}. \qquad (10.33)$$

Proof. Fix $t \in [-\pi, \pi]$ and observe, from Hardy's inequality [51] and the periodicity of u, that, as an almost-everywhere defined function of s,

$$\frac{u(t) - u(s)}{2 \tan \frac{1}{2}(t - s)}$$

is in $L_2[-\pi, \pi]$ with norm bounded by $2\|u'\|_{L_2[-\pi,\pi]}$, independent of t. Therefore

$$u(t)\mathcal{C}u'(t) - \mathcal{C}(uu')(t) = \frac{1}{2\pi} \int_{-\pi}^{\pi} \frac{u(t) - u(s)}{\tan \frac{1}{2}(t - s)} u'(s) ds$$

$$\le \frac{2}{\pi} \|u'\|^2_{L_2[-\pi,\pi]}.$$

Let $t \in \mathbb{R}$ be such that $s \mapsto \int_t^s u'(x)dx$ is differentiable with respect to s at $s = t$. According to Lebesgue's theorem [51] this set has full measure in $\mathbb{R}$. For almost all such t,

$$u(t)\mathcal{C}u'(t) - \mathcal{C}(uu')(t) = \frac{1}{2\pi}\int_{-\pi}^{\pi}(u(t) - u(s))u'(s)\cot(\tfrac{1}{2}(t-s))ds$$

$$= \frac{-1}{4\pi}\int_{-\pi}^{\pi}\cot(\tfrac{1}{2}(t-s))\frac{d}{ds}\left(\int_{t}^{s}u'(x)dx\right)^{2}ds$$

$$= \frac{1}{8\pi}\int_{-\pi}^{\pi}\left\{\frac{u(t)-u(s)}{\sin\tfrac{1}{2}(t-s)}\right\}^{2}ds \geq 0.$$

$\square$

The equality in (10.33) holds for all $t \in \mathbb{R}$ when u is 2π-periodic and infinitely differentiable. Therefore, for such functions, integrating both sides of (10.33) and Parseval's identity gives

$$2\pi\sum_{k\in\mathbb{Z}}|k||\hat{u}(k)|^{2} = \int_{-\pi}^{\pi}\frac{1}{8\pi}\int_{-\pi}^{\pi}\left\{\frac{u(t)-u(s)}{\sin\tfrac{1}{2}(t-s)}\right\}^{2}dsdt.$$

LEMMA 10.5.2 *Suppose* $u' \in L_{p}[-\pi,\pi]$ *for some* p *with* $2 < p < \infty$. *Then* $\mathcal{Q}(u) \in C^{1-\frac{2}{p}}$.

Proof. Since u is periodic and $\mathcal{Q}$ commutes with translations, it will suffice to show that $|\mathcal{Q}(u)(x) - \mathcal{Q}(u)(0)| \leq$ const $|x|^{1-\frac{2}{p}}$, for all x with $|x| < \pi/2$, where the constant depends only on $\|u'\|_{p}$ and on p. Without loss of generality suppose $x > 0$. Let $\frac{1}{p} + \frac{1}{q} = 1$ and let I denote the set $[-\pi,\pi] \setminus [-2x, 2x]$. Then

$$|\mathcal{Q}(u)(x) - \mathcal{Q}(u)(0)|$$

$$\leq \frac{1}{8\pi}\int_{-2x}^{2x}\frac{\left\{\int_{x-y}^{x}u'(t)dt\right\}^{2}}{\sin^{2}(y/2)}dy + \frac{1}{8\pi}\int_{-2x}^{2x}\frac{\left\{\int_{-y}^{0}u'(t)dt\right\}^{2}}{\sin^{2}(y/2)}dy$$

$$+ \left|\frac{1}{8\pi}\int_{I}\frac{\left\{\int_{x-y}^{x}u'(t)dt\right\}^{2} - \left\{\int_{-y}^{0}u'(t)dt\right\}^{2}}{\sin^{2}(y/2)}dy\right|.$$

Since $u' \in L_{p}[-\pi,\pi]$, $p > 2$, it follows from Hölder's inequality that

$$\int_{-2x}^{2x}\frac{\left\{\int_{x-y}^{x}u'(t)dt\right\}^{2}}{\sin^{2}(y/2)}dy + \int_{-2x}^{2x}\frac{\left\{\int_{-y}^{0}u'(t)dt\right\}^{2}}{\sin^{2}(y/2)}dy$$

$$\leq \text{ const }\|u'\|_{p}^{2}|x|^{\frac{2}{q}-1},$$

where the constant depends only on p. Also,

$$\left| \int_I \frac{\left\{ \int_{x-y}^x u'(t)dt \right\}^2 - \left\{ \int_{-y}^0 u'(t)dt \right\}^2}{\sin^2(y/2)} dy \right| \leq$$

$$\left| \int_I \frac{\left\{ \int_{x-y}^x u'(t)dt + \int_{-y}^0 u'(t)dt \right\} \left\{ \int_{x-y}^x u'(t)dt - \int_{-y}^0 u'(t)dt \right\}}{\sin^2(y/2)} dy \right|$$

$$\leq \text{const } \|u'\|_p \int_I \frac{|y|^{\frac{1}{q}} \left| \int_{x-y}^x u'(t)dt - \int_{-y}^0 u'(t)dt \right|}{\sin^2(y/2)} dy$$

$$= \text{const } \|u'\|_p \int_I \frac{|y|^{\frac{1}{q}} \left| \int_0^x u'(t)dt - \int_{-y}^{x-y} u'(t)dt \right|}{\sin^2(y/2)} dy$$

$$\leq \text{const } \|u'\|_p^2 |x|^{\frac{1}{q}} \int_I |y|^{\frac{1}{q}-2} dy \leq \text{const } \|u'\|_p^2 |x|^{\frac{2}{q}-1}.$$

Thus

$$|\mathcal{Q}(u)(x) - \mathcal{Q}(u)(0)| \leq \text{const } \|u'\|_p^2 |x|^{\frac{2}{q}-1} = \text{const } \|u'\|_p^2 |x|^{1-\frac{2}{p}}.$$

This completes the proof. $\square$

LEMMA 10.5.3 *Suppose that $u' \in C^\alpha$, $\alpha \in (0,1)$. Then $\mathcal{Q}(u) \in C^{1,\delta}$, $0 < \delta < \alpha$.*

Proof. We use the notation of the preceding proof. Since $u' \in C^\alpha$, it follows from first principles and the dominated convergence theorem that $\mathcal{Q}(u)$ is differentiable, and its derivative is given by

$$\mathcal{Q}(u)'(x) = \frac{1}{4\pi} \int_{-\pi}^\pi \frac{(u'(x) - u'(x-y)) \int_{x-y}^x u'(t)dt}{\sin^2(y/2)} dy.$$

To see that $\mathcal{Q}(u)'$ is in C^δ, $0 < \delta < \alpha$, it suffices, as in the proof of Lemma 10.5.2, to show that $|\mathcal{Q}(u)'(x) - \mathcal{Q}(u)'(0)| \leq \text{const } |x|^\delta$, where the constant depends only on α, δ and on $\|u'\|_{C^\alpha}$. Let $I = [-\pi, \pi] \setminus [-2x, 2x]$ with $x \in (0, \pi/2)$. Note first that

$$\int_{-2x}^{2x} \left| \frac{u'(x) - u'(x-y)}{\sin^2(y/2)} \int_{x-y}^x u'(t)dt \right| dy \leq \text{const } |x|^\alpha,$$

because $g \in C^\alpha$, and similarly that

$$\int_{-2x}^{2x} \left| \frac{u'(0) - u'(0-y)}{\sin^2(y/2)} \int_{-y}^0 u'(t)dt \right| dy \leq \text{const } |x|^\alpha.$$

Next note from the definition of I and the fact that $u' \in C^\alpha$,

$$\left| \int_I \frac{(u'(x) - u'(x-y))}{\sin^2(y/2)} \int_{x-y}^x u'(t)dtdy \right.$$

$$\left. - \int_I \frac{(u'(0) - u'(-y))}{\sin^2(y/2)} \int_{-y}^0 u'(t)dtdy \right|$$

$$\leq \int_I \left| \frac{(u'(x) - u'(0)) - (u'(x-y) - u'(-y))}{\sin^2(y/2)} \int_{x-y}^x u'(t)dt \right| dy$$

$$+ \int_I \left| \frac{u'(0) - u'(-y)}{\sin^2(y/2)} \left(\int_{x-y}^x u'(t)dt - \int_{-y}^0 u'(t)dt \right) \right| dy$$

$$\leq \text{const } |x|^\alpha \log|x|.$$

Combining these estimates gives that $\mathcal{Q}(u)' \in C^\delta, 0 < \delta < \alpha$, as required. $\qquad\square$

Regularity

Here are some properties of $\mathcal{Q}$ which follow from §10.5 and the theorems of Privalov and M. Riesz. Suppose that $u' \in L_2[-\pi, \pi]$ and recall the definition of G and $\mathcal{U}$ from (10.20) and (10.21).

(a) $\mathcal{Q}(u)(x) > 0$ for almost all $x \in [-\pi, \pi]$ unless $u \equiv$ constant.
(b) $\mathcal{Q}(u) \in L_\infty[-\pi, \pi]$.
(c) $\mathcal{Q}(u) \in C^\alpha$ for all $\alpha \in (0,1)$ if $u' \in L_p[-\pi, \pi]$ for all $p < \infty$.
(d) $\mathcal{Q}(u) \in C^{1,\alpha}$ for all $\alpha \in (0,1)$ if $u' \in C^\beta$ for all $\beta \in (0,1)$.

THEOREM 10.5.4 *Suppose that* $(\lambda, w) \in \mathcal{U}$ *is a solution of* (10.16). *Then* $w \in C^{2,\alpha}$.

Proof. This follows by a simple bootstrap argument. Assume that $1 - 2\lambda w > 0$ and w satisfies (10.16). Then (10.18) and (b) imply that

$$\mathcal{C}w' = \frac{\lambda(w - \mathcal{Q}(w))}{1 - 2\lambda w} \in L_\infty[-\pi, \pi] \subset L_p[-\pi, \pi] \qquad (10.34)$$

for all $p \geq 1$. Hence, by Riesz's theorem, $w' \in L_p[-\pi, \pi], 1 < p < \infty$, and so $w \in C^\beta$ for all $\beta \in (0,1)$, by Hölder's inequality, and $\mathcal{Q}(w) \in C^\beta$ for all $\beta \in (0,1)$ by (c). Therefore, by (10.34), $\mathcal{C}w' \in C^\beta$ for all $\beta \in (0,1)$ from which it follows, by Privalov's theorem, that $w' \in C^\beta, \beta \in (0,1)$. Therefore $\mathcal{Q}(w) \in C^{1,\alpha}$, by (d) and $1 - 2\lambda w \in C^{1,\alpha}$. Therefore, by (10.34), $\mathcal{C}w' \in C^{1,\alpha}$ for all $\alpha \in (0,1)$. It now follows, by Privalov's theorem, that $w \in C^{2,\alpha}$ for all $\alpha \in (0,1)$. This completes the proof. $\qquad\square$

It is clear that this procedure could be continued indefinitely to prove by induction that w can be extended as infinitely differentiable 2π-periodic function on $\mathbb{R}$. Theorem 10.3.2 (b) shows that the hypothesis $1 - 2\lambda w > 0$ (strict inequality) everywhere is essential to this conclusion.

REMARK 10.5.5 If $\{(\lambda_k, w_k)\}$ is a sequence of solutions of (10.16) with the property that $1 - 2\lambda w_k(t) \geq d > 0$, $t \in [-\pi, \pi]$ and $\{\lambda_k\}$ is bounded, then (10.12) gives that $\{w'_k\}$ is bounded in $L_2[-\pi, \pi]$ and hence, by the preceding argument, $\{w_k\}$ is bounded in $C^{2,\alpha}$ for all $\alpha \in (0, 1)$. Therefore, by the last sentence in Definition 10.1.1, $\{w_k\}$ is compact in C^2.   □

Fredholm Property

We begin with a general observation.

THEOREM 10.5.6 $\mathcal{Q}$ *is sequentially continuous from* Y, *with the weak topology, into* $L_p[-\pi, \pi]$, $1 < p < \infty$, *with the strong* L_p-*topology.*

Proof. Let $\{v_n\}$ be a sequence in Y which is weakly convergent to v. Then $v'_n \in Z$ and $v'_n \rightharpoonup v'$ in $L_2[-\pi, \pi]$. It will suffice to show for a subsequence that $\mathcal{Q}(v_n) \to \mathcal{Q}(v)$ in $L_p[-\pi, \pi]$, $1 < p < \infty$, as $n \to \infty$. Let $u_n(t) = \int_0^t (v'_n(s) - v'(s))ds$. Since $\mathcal{Q}(u_n)(t) = \mathcal{Q}(v_n - v)(t)$ and since the integral of $\mathcal{C}u$ is zero for any $u \in L_2[-\pi, \pi]$,

$$\int_{-\pi}^{\pi} \mathcal{Q}(v_n - v)(t)dt = \int_{-\pi}^{\pi} u_n(t)\mathcal{C}(u_n{}')(t)dt$$

$$= \int_{-\pi}^{\pi} u_n(t)\mathcal{C}(v'_n - v')(t)dt \to 0$$

as $n \to \infty$, because $\mathcal{C}$ is bounded on $L_2[-\pi, \pi]$, $v'_n \rightharpoonup v'$ and $u_n \to 0$ in $L_2[-\pi, \pi]$.

Since $\mathcal{Q}(v_n - v)$ is non-negative, it follows that $\mathcal{Q}(v_n - v) \to 0$ in $L_1[-\pi, \pi]$ and so a subsequence converges pointwise almost everywhere. Let

$$\phi_n^t(s) = \frac{\int_{t-s}^t v'_n(t)dt}{\sin(s/2)} \text{ and } \phi^t(s) = \frac{\int_{t-s}^t v'(t)dt}{\sin(s/2)}.$$

We have just shown for a subsequence that, for almost all t, $\phi_n^t \to \phi^t$ in $L_2[-\pi, \pi]$ as $n \to \infty$, and hence that $\|\phi_n^t\|_{L_2[-\pi, \pi]} \to \|\phi^t\|_{L_2[-\pi, \pi]}$. Another way of saying this is that, for almost all t, $\mathcal{Q}(v_n)(t) \to \mathcal{Q}(v)(t)$ as $n \to \infty$. Since $\{\mathcal{Q}(v_n)\}$ is bounded in $L_\infty[-\pi, \pi]$, by Lemma 10.5.1, the result of the theorem follows from the dominated convergence theorem.   □

To show that the linearization of G at $(\lambda, w) \in \mathcal{U}$ is Fredholm with index 0 it suffices (Theorem 2.7.6) to decompose $\partial_w G[(\lambda, w)]$ as

$$\partial_w G[(\lambda, w)] = K[\lambda, w] + C[\lambda, w],$$

where $K[\lambda, w] : Y \to X$ is a homeomorphism and $C[\lambda, w] : Y \to X$ is compact. Since, for $(\lambda, w) \in \mathbb{R} \times Y$,

$$G(\lambda, w) = (1 - 2\lambda w)(\mathcal{C}w' + w) + \lambda \mathcal{Q}(w) - (1 - 2\lambda w + \lambda)w,$$

differentiation gives

$$\partial_w G[(\lambda, w)]\varphi = (1 - 2\lambda w)(\mathcal{C}\varphi' + \varphi)$$
$$+ \left(\lambda d\mathcal{Q}[w]\varphi - 2\lambda \varphi (\mathcal{C}w' + w) - \lambda \varphi + 4\lambda w \varphi - \varphi \right). \quad (10.35)$$

Since $1 - 2\lambda w$ is everywhere positive and continuous, the first term on the right obviously represents a homeomorphism from Y to X. Since the nonlinear operator $\mathcal{Q}$ is compact from Y into X (Theorem 10.5.6), its Fréchet derivative $d\mathcal{Q}[w] : Y \to X$ is also compact (Lemma 3.1.12). Also since, by the Ascoli-Arzelà theorem 2.7.1, Y is compactly embedded (see Example 2.7.3) in $L_\infty[-\pi, \pi]$, all the other terms in the second bracket of (10.35) denote compact linear operators from Y into X.

This shows that $\partial_w G[(\lambda, w)] : Y \to X$ is Fredholm with index 0 on $\mathcal{U}$. Note that $\mathcal{U}$ is open in $\mathbb{R} \times Y$.

10.6 NOTES ON SOURCES

For the maximum principle, the Hopf boundary-point lemma and the Phragèmen-Lindelöf principle, see [31] and [49]. See also the Notes on Sources for Chapter 11.

Chapter Eleven

Global Existence of Stokes Waves

11.1 LOCAL BIFURCATION THEORY

Here we study the existence of non-trivial solutions of equation (10.16) in a neighbourhood of the trivial solution $w = 0 \in Y$ for values of $\lambda > 0$. This equation can be re-written as $G(\lambda, w) = 0$ where G, defined by (10.20), has the following properties.

(A) $G(\lambda, 0) = 0$ for all $\lambda > 0$.

(B) $\partial_w G[(\lambda, w)]\varphi = \mathcal{C}\varphi' - \lambda\varphi - \lambda(w\mathcal{C}\varphi' + \varphi\mathcal{C}w' + \mathcal{C}(w\varphi)')$.

(C) $\partial_\lambda G[(\lambda, w)]\mu = \mu(w + w\mathcal{C}w' + \mathcal{C}(ww'))$.

Since $\partial_w G : (0, \infty) \times Y \to \mathcal{L}(Y, X)$ and $\partial_\lambda G : (0, \infty) \times Y \to \mathcal{L}(\mathbb{R}, X)$ are both continuous, we conclude (Lemma 3.1.7) that $G : (0, \infty) \times Y \to X$ is continuously Fréchet differentiable. In fact all its partial Fréchet derivatives of all orders greater than three are zero everywhere. Thus $G : (0, \infty) \times Y \to X$ is a real-analytic operator.

Since G is continuously Fréchet differentiable and $\partial_w G[(\lambda, 0)]$ is Fredholm with index zero, a necessary condition (see Example 8.1.1) for there to be bifurcation from the trivial solution at $\lambda_0 >$ is that

$$\partial_w G[(\lambda_0, 0)] : Y \to X \text{ is not injective,}$$

equivalently

$$\mathcal{C}\varphi' - \lambda_0\varphi = 0 \text{ has a non-trivial solution in } Y.$$

Bifurcation from a Simple Eigenvalue

Since

$$\partial_w G[(\lambda, 0)]\varphi = \mathcal{C}\varphi' - \lambda\varphi = 0, \ \varphi \in Y \setminus \{0\},$$

if and only if $n \in \mathbb{N}_0$ and $\varphi(t) = a\cos nt$, $a \in \mathbb{R}$, all bifurcation points λ_0 are non-negative integers. Now we will show that every non-negative integer is a bifurcation point.

It is easily seen that $\lambda_0 = 0$ is a bifurcation point since (10.16) has a solution $(\lambda, w) = (0, c)$ for all constants c. Now suppose that $\lambda_0 = n \in \mathbb{N}$. We need only check the hypotheses of the theory of bifurcation from a simple eigenvalue. Note that

$$\ker \partial_w G[(n, 0)] = \text{span} \{\varphi_n\} \text{ and } \varphi_n \notin \text{range } \partial_w G[(n, 0)],$$

where $\varphi_n(t) = \cos nt$. Thus every $n \in \mathbb{N}$ is a simple eigenvalue of $A : Y \to X$, where $Au = Cu'$, and hence every $n \in \mathbb{N}_0$ is a bifurcation point for the equation $G(\lambda, w) = 0$.

THEOREM 11.1.1   *Let $n \in \mathbb{N}_0$. Then there is a neighbourhood O_n of $(n, 0)$ in $\mathbb{R} \times Y$, $\epsilon > 0$ and a unique real-analytic function $(\Lambda_n, \Phi_n) : (-\epsilon, \epsilon) \to O_n$ with*

$$\int_{-\pi}^{\pi} \varphi_n(t)\Phi_n(s)(t)dt = 0, \quad s \in (-\epsilon, \epsilon),$$

such that $(\lambda, w) \in O_n$ is a solution of (10.16) with $w \neq 0$ if and only in there exists $s \in (-\epsilon, \epsilon) \setminus \{0\}$ such that

$$(\lambda, w) = (\Lambda_n(s), s(\varphi_n + \Phi_n(s))) \ where \ (\Lambda_n(0), \Phi_n(0)) = (n, 0).$$

Although the bifurcation from $\lambda_0 = 0$ was known from the elementary observation that constants are solutions when $\lambda = 0$, the theorem gives the additional information that in a neighbourhood of $(0, 0)$ in $\mathbb{R} \times Y$ there are no other solutions.

Bifurcation from $\lambda = 1$

Let $\mathcal{E}$ denote the set of all non-trivial solutions $(\lambda, w) \in (0, \infty) \times Y$ of (10.16) and let

$$\Gamma_1 = \{(\Lambda_1(s), s(\varphi_1 + \Phi_1(s))) : s \in (-\epsilon, \epsilon) \setminus \{0\}\} \subset \mathcal{E}$$

denote the curve of non-trivial solutions bifurcating from $(1, 0)$. The theory so far predicts that in a neighbourhood of the bifurcation point $(1, 0)$ there are no non-trivial solutions of (10.16) other than those on the bifurcating curve Γ_1. Let $\mathcal{E}_1$ denote the maximal connected subset in $\mathbb{R} \times Y$ of $\overline{\mathcal{E}}$ which contains Γ_1. Now observe that in a neighbourhood of $(1, 0)$ the curve Γ_1 is symmetric in $\mathbb{R} \times Y$ about the λ-axis because $(\lambda, \tilde{w}) \in \mathcal{E}_1$ whenever $(\lambda, w) \in \Gamma_1$ (see (10.15)) and, by the uniqueness of the function (Λ_1, Φ_1), since $\tilde{\varphi}_1 = -\varphi_1$,

$$(\Lambda_1(-s), (-s)(\varphi_1 + \Phi_1(-s))) = (\Lambda_1(s), s(\varphi_1 + \Phi_1(s))^{\sim})$$

for all $s \in (0, \epsilon)$. Hence $\dot{\Lambda}_1 = (d/ds)\Lambda_1(s)|_{s=0} = 0$. (Here $\dot{}$ denotes differentiation with respect to s at $s = 0$.) In other words the tangent to the bifurcating curve Γ_1 is vertical in $\mathbb{R} \times Y$ at the bifurcation point. We shall show presently that the whole curve Γ_1 is not vertical, unlike the case of bifurcation from $\lambda_0 = 0$. Let

$$\Gamma_1^+ := \{(\Lambda_1(s), s(\varphi_1 + \Phi_1(s))) : s \in (0, \epsilon)\}, \quad Y_1 = \text{span}\{\varphi_1\}$$

and let Y_2 denote the closure in Y of $\text{span}\{\varphi_k : k = 0, 2, 3, \dots\}$.

Substituting $(\lambda, w) = (\Lambda_1(s), s(\varphi_1 + s\Phi_1(s)))$ in (10.16) gives

$$\begin{aligned} 0 = (1 - \Lambda_1(s))\varphi_1 + &\mathcal{C}\Phi_1(s)' - \Lambda_1(s)\Phi_1(s) \\ &- s\Lambda_1(s)\{(\varphi_1 + \Phi_1(s))\mathcal{C}(\varphi_1' + \Phi_1(s)') \\ &\qquad + \mathcal{C}((\varphi_1 + \Phi_1(s))(\varphi_1 + \Phi_1(s))')\}. \quad (11.1) \end{aligned}$$

A differentiation with respect to s at $s = 0$, using the fact that $\dot{\Lambda}_1 = 0$, gives

$$C\dot{\Phi}'_1 - \dot{\Phi}_1 = \varphi_1 C\varphi'_1 + C(\varphi_1 \varphi'_1) = \varphi_1^2 + \frac{1}{2}\varphi_2 = \frac{1}{2} + \varphi_2.$$

Therefore $\dot{\Phi}_1 = \varphi_2 - \frac{1}{2}$. Differentiating (11.1) again with respect to s at $s = 0$ gives

$$0 = -\ddot{\Lambda}_1 \varphi_1 + C\ddot{\Phi}'_1 - \ddot{\Phi}_1 - 2\varphi_1 - 6\varphi_3.$$

Hence

$$\ddot{\Lambda}_1 = -2 \text{ and } \ddot{\Phi}_1 = 3\varphi_3.$$

Therefore Γ_1 is a symmetric, sub-critical pitchfork bifurcation at $(1,0)$.

Consider the eigenvalue problem for the linearization of (10.16) with respect to w at $(\Lambda_1(s), w_s)$, where $w_s = s(\varphi_1 + \Phi_1(s))$,

$$\partial_w G[(\Lambda_1(s), w_s)]v = \nu v, \quad v \in X. \tag{11.2}$$

In the notation of Proposition 8.4.2, $F = -G$ (see (8.4) and (10.20)) and so $\nu = -\mu$. Therefore, for s sufficiently small, all solutions (ν, v) of (11.2) in a sufficiently small neighbourhood of $(0, \varphi_1)$ (up to a normalization of v) are given by a smooth mapping $s \mapsto (\nu_s, v_s)$ such that $(\nu_0, v_0) = (0, \varphi_1)$ and

$$\lim_{s \to 0} \frac{sd\Lambda_1(s)/ds}{\nu_s} = 1.$$

Further, from the symmetry of Γ_1 we may infer that ν_s is even in s and $v_{-s} = \tilde{v}_s$. These conclusions may be summarized as follows:

$$\left. \begin{array}{rcl} w_s := s(\varphi_1 + \Phi_1(s)) &=& s\varphi_1 + s^2(\varphi_2 - \frac{1}{2}) + \frac{3}{2}s^3\varphi_3 \\ \Lambda_1(s) &=& 1 - s^2 \\ \nu_s &=& -2s^2 \end{array} \right\} + O(|s|^4). \tag{11.3}$$

as $|s| \to 0$. For non-zero s with $|s|$ sufficiently small, all but two of the eigenvalues of $\partial_w G[(\Lambda_1(s), w_s)]$ are greater than 1/2, since that is true when $s = 0$ and the eigenvalues ν of (11.2) depend continuously on s. Of the two exceptions, one is close to -1 (because -1 is an eigenvalue of (11.2) when $s = 0$), and the other is negative by (11.3). Thus, when positive $|s|$ is sufficiently small, equation (11.2) has exactly two negative eigenvalues and all the others exceed 1/2. We return briefly to this observation after we have defined Morse indices in §11.3: it says that near the bifurcation point $\lambda = 1$ all non-trivial solutions have Morse index 2.

11.2 GLOBAL BIFURCATION FROM $\lambda = 1$

Now we turn to a global analysis of the water-wave problem, exploiting its real analyticity and the Fredholm property §10.5. In the same notation, let $\mathcal{E}_1^+$ denote the closure of the maximal connected subset of $\mathcal{E} \cap \mathcal{U}$ in $(0, \infty) \times Y$ which contains Γ_1^+ ($\mathcal{U}$ is defined in (10.21)). The outcome real-analytic bifurcation theory is the following result on the existence of a global continuum in $\mathcal{U}$ of solutions of (10.16). Let

$$S = \{(\lambda, w) \in \mathcal{E}_1^+ : \partial_w G[(\lambda, w)] : Y \to X \text{ is a homeomorphism.}\}$$

THEOREM 11.2.1 *For equation* (10.16) *there exists a function* $h = (\Lambda, W) :$ $(0, \infty) \to \mathbb{R} \times Y$ *such that*

(i) $h : (0, \infty) \to \bar{S} \cap \mathcal{U}$ *is continuous;*

(ii) $h((0, 1)) \subset \Gamma_1^+$ *and* $\lim_{s \searrow 0} h(s) = (1, 0)$;

(iii) h *is injective on* $h^{-1}(S)$;

(iv) *at all points* $s \in h^{-1}(S)$, *h is real-analytic with* $\Lambda'(s) \neq 0$;

(v) *the set* $h^{-1}(\bar{S} \backslash S) \subset (0, \infty)$ *consists of isolated values;*

(vi) *locally near every point* $s_0 \in h^{-1}(\bar{S} \backslash S)$, *there exists an injective and continuous re-parameterization of the set originally parameterized by h, such that* $s = \gamma(\sigma)$, $|\sigma| \leq 1$, $s_0 = \gamma(0)$ *and* $h \circ \gamma$ *is a real-analytic function (whose derivative may vanish only at 0).*

Let $\mathfrak{R} = \{h(s) : s \in [0, \infty)\}$. Since $\mathfrak{R} \subset \mathcal{U}$ and $\mathfrak{R}$ is path-connected in Y, it follows from the remark at the end of §10.5 that $\mathfrak{R}$ is path-connected in C^k, for all $k \in \mathbb{N}$. Let

$$\mathcal{W} = \Big\{(\lambda, w) \in \mathbb{R} \times C^{2,\alpha} \text{which satisfies (10.16) and}$$

$$1 + \mathcal{C}w' > 0 \text{ on } [-\pi, \pi], \quad w' < 0 \text{ on } (0, \pi), \quad w''(0) < 0 < w''(\pi)\Big\}.$$

It is easy to see from the local bifurcation analysis of the preceding section that $h(s) \in \mathcal{W}$ for $s \in (0, \epsilon)$ when $\epsilon > 0$ is sufficiently small. However Lemma 10.4.1 implies that $\mathcal{W} \cap \mathfrak{R}$ is both open and closed in $\mathfrak{R}$ and hence, by path-connectedness, $\mathfrak{R} \subset \mathcal{W}$. From this and Theorem 10.4.4 it follows that if $(\lambda, w) \in \mathfrak{R}$, then $0 < k \leq \lambda \leq K < \infty$ and $|w| \leq K$, for some $k, K > 0$. Now it follows from (10.12) that if $1 - 2\lambda w(0) \geq \delta > 0$ for all $(\lambda, w) \in \mathfrak{R}$, then $\mathfrak{R}$ is bounded in $\mathbb{R} \times Y$. It must therefore form a closed loop, by Theorem 9.1.1 (e), if $1 - 2\lambda w(0) \geq \delta > 0$ for all $(\lambda, w) \in \mathfrak{R}$.

Now let

$$\mathcal{K} = \{w \in Y : w' \leq 0 \text{ on } (0, \pi)\}.$$

We have seen that $\mathfrak{R} \subset \mathcal{K}$ and $\mathcal{K}$ is a closed cone in Y. In order to show that $\mathfrak{R}$ is not a closed loop in Y it suffices to confirm the hypotheses of Theorem 9.2.2. It has already been observed that hypotheses (a) and (b) hold, and (c) holds also because

$\xi_0 = \cos t \in \mathcal{K}$. Suppose that $(\lambda, w) \in \mathfrak{R}$. Then $w \in C^{2,\alpha}$ and $\vartheta \leq \pi/3$ on $(0, \pi)$. Moreover all solutions of (10.16) close to (λ, w) in $\mathbb{R} \times Y$ are close in $\mathbb{R} \times C^{2,\alpha}$ and the corresponding $\vartheta \leq \pi/3$. This proves that hypothesis (d) holds and so $\mathfrak{R}$ is not a closed loop.

We conclude that

$$h(s) \to \partial \mathcal{U} \text{ as } s \to \infty,$$

$\mathfrak{R}$ lies in a bounded subset of $[k, \infty) \times W^{1,p}[-\pi, \pi]$, for some $k > 0$ and any $p \in [1, 3)$ (Theorem 10.4.4), and

$$1 - 2\lambda w(0) \to 0 \text{ as } s \to \infty \text{ when } h(s) = (\lambda, w).$$

Here $W^{1,p}[-\pi, \pi]$ denotes the Banach space of 2π-periodic functions with weak derivative in $L_p[-\pi, \pi]$. It follows that every sequence in $\mathfrak{R}$ has a subsequence which is convergent weakly in $[k, \infty) \times W^{1,p}[-\pi, \pi]$, $1 < p < 3$, and strongly in $[k, \infty) \times C^\alpha$, $\alpha \in (0, 2/3)$. Let

$$\mathcal{H} = \overline{\mathfrak{R}} \setminus \mathfrak{R} \text{ in } C^{\frac{1}{2}}.$$

LEMMA 11.2.2 $\mathcal{H} \subset \mathbb{R} \times W^{1,p}[-\pi, \pi]$, $1 < p < 3$ and every $(\lambda, w) \in \mathcal{H}$ satisfies (10.16) with $1 - 2\lambda w(0) = 0$. Moreover the set $\mathcal{H}$ is a compact, connected subset of $\mathbb{R} \times C^{\frac{1}{2}}$.

Proof. Let $\{h(s_k)\} = \{(\lambda_k, w_k)\}$ be a sequence in $\mathfrak{R}$ which converges to $(\lambda, w) \in \mathcal{H}$ in $C^{\frac{1}{2}}$ as k and $s_k \to \infty$. Extracting a subsequence if necessary, we may suppose that it also converges weakly to (λ, w) in $W^{1,p}[-\pi, \pi]$. Thus $(\lambda, w) \in \mathbb{R} \times W^{1,p}[-\pi, \pi]$. Multiplying (10.16) by a smooth 2π-periodic function ϕ and integrating before taking the limit gives

$$0 = \int_{-\pi}^{\pi} \phi \left\{ \mathcal{C}w_k' - \lambda_k(w_k + w_k \mathcal{C}w_k' + \mathcal{C}(w_k w_k')) \right\} dt$$
$$\to \int_{-\pi}^{\pi} \phi \left\{ \mathcal{C}w' - \lambda(w + w\mathcal{C}w' + \mathcal{C}(ww')) \right\}$$

because $\mathcal{C}w_k' \to \mathcal{C}w'$, $w_k' \to w'$ weakly in Y and $w_k \to w$ uniformly on $[-\pi, \pi]$ as $k \to \infty$. It follows that $(\lambda, w) \in \mathbb{R} \times W^{1,p}[-\pi, \pi]$ satisfies (10.16).

It follows that $\mathcal{H} = \overline{\mathfrak{R}} \setminus \mathfrak{R}$ is compact in $\mathbb{R} \times C^{\frac{1}{2}}$. To show that $\mathcal{H}$ is connected in $\mathbb{R} \times C^{\frac{1}{2}}$, suppose that it is not. Then there exists two disjoint compact (in $\mathbb{R} \times C^{\frac{1}{2}}$) subsets U, V such that both intersect $\mathcal{H}$ non-trivially, their union is $\mathcal{H}$, and $U \cap V$ is empty. Let $\text{dist}(U, V) = d > 0$. Suppose that $u \in \mathcal{H} \cap U$ and $v \in \mathcal{H} \cap V$. Since $u \notin V$ and $v \notin U$, there exists $h(t_m)$ with $\text{dist}(h(t_m), U) \geq d/4$, $\text{dist}(h(t_m), V) \geq d/4$ and $t_m \to \infty$ as $m \to \infty$. The corresponding sequence $\{(\lambda_m, w_m)\}$ has a subsequence which converges strongly in $C^{\frac{1}{2}}$ to a solution (λ, w) of (10.16) with $1 - 2\lambda w(0) = 0$ and $\text{dist}((\lambda, w), U) \geq d/4$, $\text{dist}((\lambda, w), V) \geq d/4$. This contradicts the fact that $(\lambda, w) \in \mathcal{H}$ and the proof is complete. $\square$

REMARK 11.2.3   The set $\mathcal{H}$ consists of Stokes waves of greatest height which arise as the limit as $s \to \infty$ of points on $\mathfrak{R}$. While it may not be a singleton, it is connected. Such a conclusion could not be derived using topological bifurcation theory because in that case there is no analogue of $h(s) \to \partial \mathcal{U}$ as $s \to \infty$, which is the key here.   □

REMARK 11.2.4   **(Scaling)** Let

$$\mathfrak{R}_k = \{(k\lambda, \vartheta(kt)) : (\lambda, \vartheta) \in \mathfrak{R}\}, \quad k \in \mathbb{N}.$$

It is easy to see that elements of $\mathfrak{R}_k$ are solutions to (10.16) that give rise to Stokes waves of period 2π, but of minimal period $2\pi/k$. The curve $\mathfrak{R}_k$ represents a global curve bifurcating at the point $(k, 0)$. It is clear from the real-analytic theory developed here that $\mathfrak{R}_k$ is in fact the same curve as would be given by a study of the global bifurcation from a simple eigenvalue at $\lambda = k$.   □

11.3 GRADIENTS, MORSE INDEX AND BIFURCATION

To finish, we give a brief sketch of how real-analytic bifurcation theory interacts with the gradient structure of (10.16) to conclude the existence of multiple secondary bifurcation points on the global branch. As the abstract theory upon which these conclusions are based is far removed from considerations of real-analyticity, we omit the proofs. It is worth emphasizing however that the existence of a path, not just a connected set, of solutions is essential. This is where the real-analyticity comes in.

Recall the notion of gradient structure in a Banach space setting (Definitions 3.4.1 and 3.4.2). For certain operators with gradient structure, the Morse index of a solution is defined as follows.

DEFINITION   *Suppose that Y is dense in a Hilbert space $(X, \langle \cdot, \cdot \rangle)$ and that $G(\lambda, \cdot)$ is the gradient of a C^2-functional $g(\lambda, \cdot)$ with $G(\lambda, y) = 0$ and $\partial_y G[(\lambda, y)] - \mu\iota : Y \to X$ a homeomorphism except for μ in a discrete set $\mathbb{S}(\lambda, y)$. For $\mu \in \mathbb{S}(\lambda, y)$ suppose that $\big(\mu\iota - \partial_y G[(\lambda, y)]\big)$ is a Fredholm operator of index zero.*

Then the Morse index $M(\lambda, y)$ of the solution (λ, y) is the number of strictly negative eigenvalues of $\partial_y G[(\lambda, y_0)]$, counted according to multiplicity. (The multiplicity of μ is $\dim \ker \big(\mu\iota - \partial_y G[(\lambda, y)]\big)$.)

DEFINITION   *We say that (λ_0, y_0) is a bifurcation point for the equation $G(\lambda, y) = 0$ if there are two sequences $\{(\lambda_k, \widehat{y}_k)\}$, $\{(\lambda_k, \widetilde{y}_k)\}$ of solutions of $G(\lambda, y) = 0$ with $\widehat{y}_k \neq \widetilde{y}_k$ for all k (note the same λ_k for both) converging to (λ_0, y_0) in $\mathbb{R} \times Y$.*

The following theorem is due to Kielhöfer [37] and independently to Chow and Lauterbach [20].

PROPOSITION   *Suppose that $\mathcal{U} \subset (0, \infty) \times Y$ is an open set, $G : \mathcal{U} \to X$ is C^2 and such that $M(\lambda, y)$ is well-defined for every $(\lambda, y) \in \mathcal{U}$ with $G(\lambda, y) = 0$.*

Suppose also that for compact sets of solutions in $\mathcal{U}$, the sets $\mathbb{S}(\lambda, y)$ are uniformly bounded below.

Let $S := \{(\lambda(s), y(s)) : s \in (-\epsilon, \epsilon)\} \subset \mathcal{U}$ be a continuous curve of solutions to $G(\lambda, y) = 0$ such that $0 \notin \mathbb{S}(\lambda(s), y(s))$ for all $s \in (-\epsilon, \epsilon) \setminus \{0\}$ and that

$$\lim_{s \nearrow 0} M(\lambda(s), y(s)) \neq \lim_{s \searrow 0} M(\lambda(s), y(s)).$$

Then $(\lambda(0), y(0))$ is a bifurcation point.

Morse Index of Stokes Waves

Recall that (10.16) can be written as $G(\lambda, w) = 0$, where $G : \mathbb{R} \times Y \to X$ is given by

$$G(\lambda, w) = \mathcal{C}w' - \lambda\{w + w\mathcal{C}w' + \mathcal{C}(ww')\}.$$

Let $\mathcal{J} : \mathbb{R} \times Y \to \mathbb{R}$ be defined by

$$\mathcal{J}(\lambda, w) = \frac{1}{2} \int_{-\pi}^{\pi} \{w\mathcal{C}w' - \lambda w^2 (1 + \mathcal{C}w')\} dt, \quad \lambda > 0, \ w \in Y,$$

Then $\nabla \mathcal{J} = G$, $G : \mathbb{R} \times Y \to X$ and $\mathcal{J} : \mathbb{R} \times Y \to \mathbb{R}$ are real-analytic. If $(\lambda, w) \in \mathcal{U}$ and $G(\lambda, w) = 0$, then the Morse index $M(\lambda, w)$ is well-defined. Since a change in Morse index results bifurcation points on the curve $\mathfrak{R}$, the following result is of profound significance.

PROPOSITION (**Plotnikov [48]**) *If $(\lambda_k, w_k) \in \mathcal{U}$ is a sequence of solutions of (10.16) in Theorem 11.2.1 with $1 - 2\lambda_k w_k(0) \to 0$ as $k \to \infty$, then $M(\lambda_k, w_k) \to \infty$.*

The following result follows immediately from the existence of a path of solutions given by the analytic global bifurcation theorem 11.2.1. (In the topological version of global bifurcation theory it is difficult, and in general not always possible, to be sure of the existence of such a path upon which secondary bifurcation points can be identified.)

PROPOSITION (**Buffoni, Dancer and Toland [13]**) *There is an infinite set Σ of values of $s > 0$ at which $h(s) \in \bar{S} \setminus S$ is a bifurcation point for the equation $G(\lambda, w) = 0$. The set $\{h(s) \in \bar{S} \setminus S, \ s \in \Sigma\} \subset \mathcal{U}$ is infinite.*

Concluding Remarks

Numerical experiments suggest that in the physical domain the curve $\mathfrak{R}$ gives a *maximal connected set* of Stokes waves of the *fundamental period* 2π, and that no Stokes waves of period 2π bifurcate from it. Nor does it self-intersect. If this is so (and there is no proof) then the bifurcation points given by the last proposition are all turning points (and $\mathfrak{R}$ has infinitely many 'S-shapes', on a bifurcation diagram) and, after scaling (Remark 11.2.4), there are analogous bifurcation points on each of the curves $\mathfrak{R}_k$, $k \in \mathbb{N}$.

Whatever the nature of these bifurcation points, their existence implies the existence of points on the curves $\mathfrak{R}_q$, for sufficiently large prime numbers q, where solutions of minimal period $2\pi/q$ bifurcate from $\mathfrak{R}_q$. These bifurcations *cannot* be re-scaled to give bifurcations on $\mathfrak{R}$.

When these observations are interpreted in the physical domain, they correspond to *period-multiplying* bifurcations of Stokes waves. The physical waves which bifurcate have minimal period $2q\pi$ and there are infinitely many period-multiplying bifurcation points for Stokes waves in the physical domain [13].

11.4 NOTES ON SOURCES

Equation (10.16) is due to Babenko [8]. This approach to the Stokes wave problem appeared in [12, 13]. Related material is to be found in [11, 14, 48, 62]. An earlier version of global Stokes-wave theory is summarized in [60].

Bibliography

[1] L. V. Ahlfors, *Conformal Invariants: Topics in Geometric Function Theory*, McGraw-Hill, New York, 1973.

[2] A. Ambrosetti and G. Prodi, *A Primer of Nonlinear Analysis*, Cambridge University Press, Cambrige, U.K., 1993.

[3] C. J. Amick, *Bounds for water waves*, Arch. Rational Mech. Anal. **99** (1987), 91–114.

[4] C. J. Amick and J. F. Toland, *On periodic water waves and their convergence to solitary waves in the long-wave limit*, Phil. Trans. Roy. Soc. Lond. A **303** (1981), 633–669.

[5] S. S. Antman, *The Theory of Rods*, Handbuch der Physik, Vol. VIa/2, Springer, Berlin, 1972.

[6] ———, *Problems of Elasticity*, Springer, New York, 1995.

[7] A. Axler, P. Bourdon, and W. Ramey, *Harmonic Function Theory*, Springer-Verlag, New York, 1992.

[8] K. I. Babenko, *Some remarks on the theory of surface waves of finite amplitude*, Soviet Math. Doklady **35** (1987), 599–603.

[9] N. K. Bary, *A Treatise on Trigonometric Series*, Pergamon Press, Oxford, 1964.

[10] H. Brezis, *Analyse Fonctionnelle*, Masson, Paris, 1983.

[11] B. Buffoni, E. N. Dancer, and J. F. Toland, *Sur les ondes de Stokes et une conjecture de Levi-Civita*, Comptes Rendus Acad. Sci. Paris, Série I **326** (1998), 1265–1268.

[12] B. Buffoni, E. N. Dancer, and J. F. Toland, *The regularity and local bifurcation of steady periodic water waves*, Arch. Rational Mech. Anal. **152** (2000), 207–240.

[13] ———, *The sub-harmonic bifurcation of Stokes waves*, Arch. Rational Mech. Anal. **152** (2000), 241–271.

[14] B. Buffoni and J. F. Toland, *Dual free boundaries for Stokes waves*, Comptes Rendus Acad. Sci. Paris, Série I **332** (2001), 73-78.

[15] H. Cartan, *Elementary theory of analytic functions of one and several complex variables*, Editions Scientifiques Hermann, Addison Wesley, Paris, 1963.

[16] ———, *Differential Calculus*, Hermann, Paris, 1971.

[17] S. D. Chatterji, *Cours d'Analyse: Volume 2*, Presses Polytechniques et Universitaires Romandes, Lausanne, 1988.

[18] E. M. Chirka, *Complex Analytic Sets*, Kluwers Academic Publishing, Dordrecht, 1989.

[19] S.-N. Chow and J. K. Hale, *Methods of Bifurcation Theory* (Corrected Printing 1996), Springer-Verlag, New York, 1982.

[20] S.-N. Chow and R. Lauterbach, *A bifurcation theorem for critical points of variational problems*, Nonlinear Analysis TMA **12** (1988), 51–61.

[21] M. G. Crandall and P. H. Rabinowitz, *Nonlinear Sturm-Liouville eigenvalue problems and topological degree*, Jour. Math. Mechanics **90** (1970), 1083–1102.

[22] ———, *Bifurcation from a simple eigenvalue*, Jour. Funct. Anal. **8** (1971), 321–340.

[23] ———, *Bifurcation, perturbation of simple eigenvalues and linearised stability*, Arch. Rational Mech. Anal. **52** (1973), 161–180.

[24] E. N. Dancer, *Bifurcation theory for analytic operators*, Proc. Lond. Math. Soc. **XXVI** (1973), 359–384.

[25] ———, *Global solution branches for positive mappings*, Arch. Rational Mech. Anal. **52** (1973), 181–192.

[26] ———, *Global structure of the solution set of non-linear real-analytic eigenvalue problems*, Proc. Lond. Math. Soc. **XXVI** (1973), 359–384.

[27] E. B. Davies *Spectral Theory and Differential Operators*, C.U.P., 1995.

[28] J. Dieudonné, *Foundations of Modern Analysis*, Academic Press, 1960.

[29] H. Federer, *Geometric Measure Theory*, Springer-Verlag, Berlin, 1969.

[30] L. E. Fraenkel, *Formulae for higher derivatives of composite functions*, Math. Proc. Camb. Phil. Soc. **83** (1978), 159–165.

[31] ———, *An Introduction to Maximum Principles and Symmetry in Elliptic Problems*, Cambridge University Press, 2000.

[32] A. Friedman, *Foundations of Modern Analysis*, Dover, New York, 1982.

[33] D. Gilbarg and N. S. Trudinger, *Elliptic Partial Differential Equations of Second Order.*, 2nd. ed., Springer-Verlag, Berlin, 1983.

[34] M. Golubitsky and V. Guillemin, *Stable Mappings and their Singularities*, Springer, New York, 1973.

[35] T. Kato, *Perturbation Theory for Linear Operators*, 2nd ed., Springer Verlag, Berlin, 1976.

[36] G. Keady and J. Norbury, *On the existence theory for irrotational water waves*, Math. Proc. Camb. Phil. Soc. **83** (1978), 137–157.

[37] H. Kielhöfer, *A bifurcation theorem for potential operators*, J. Functional Analysis **77** (1988), 1–8.

[38] K. Knopp, *Theory and Application of Infinite Series*, Blackie & Sons, London & Glasgow, 1951.

[39] M. A. Krasnosel'skii, *Topological Methods in the Theory of Nonlinear Integral Equations*, Pergamon Press, Oxford, 1964.

[40] Yu. P. Krasov'skii, *On the theory of steady waves of finite amplitude*, U.S.S.R. Comput. Math. math. Phys. **1** (1961), 996–1018.

[41] E. Kreyszig, *Introductory Functional Analysis with Applications*, John Wiley and sons, New-York, 1978.

[42] T. Levi-Civita, *Détermination rigoureuse des ondes permanentes d'ampleur finie*, Math. Annalen **93** (1925), 264–314.

[43] S. Lojasiewicz, *Introduction to Complex Analytic Geometry*, Birkhäuser Verlag, Basel, 1991.

[44] J. B. McLeod, *The Stokes and Krasovskii conjectures for the wave of greatest height*, Studies in Applied Math. **98** (1997), 311–334, (In pre-print-form: Univ. of Wisconsin Mathematics Research Center Report Number 2041, 1979 (*sic*)).

[45] D. Mumford, *Algebraic Geometry I: Complex Projective Varieties*, Springer-Verlag, Berlin, 1976.

[46] R. Narasimhan, *Introduction to the Theory of Analytic Spaces*, Lecture Notes in Mathematics, no. 25, Springer-Verlag, Heidelberg, 1966.

[47] A. I. Nekrasov, *The exact theory of steady waves on the surface of a heavy fluid*, Izdat. Akad. Nauk. SSSR, Moscow, 1951. Translated as Univ. of Wisconsin MRC Report No.813, 1967.

[48] P. I. Plotnikov, *Nonuniqueness of solutions of the problem of solitary waves and bifurcation of critical points of smooth functionals*, Math. USSR Izvestiya **38 (2)** (1992), 333–357.

[49] M. H. Protter and H. F. Weinberger, *Maximum Principles in Differential Equations*, Prentice-Hall, New Jersey, 1967.

[50] P. H. Rabinowitz, *Some global results for nonlinear eigenvalue problems*, Jour. Funct. Anal. **7** (1971), 487–513.

[51] W. Rudin, *Real and Complex Analysis. 3rd edn.*, McGraw-Hill, New York, 1986.

[52] J. T. Schwartz, *Nonlinear Functional Analysis*, Gordon and Breach, New York, 1969.

[53] E. Shargorodsky and J. F. Toland, *A Riemann-Hilbert problem and the Bernoulli boundary condition in the variational theory of Stokes waves*, to appear in Annales Inst. H. Poincaré.

[54] C. A. Stuart, *An introduction to bifurcation theory based on differential calculus*, Nonlinear Mathematics and Mechanics: Heriot-Watt Symposium, Vol. IV (R. J. Knops, ed.), Research Notes in Mathematics, no. 39, Pitman, 1979, pp. 76–132.

[55] ———, *Théorie de la Bifurcation*, cours polycopié, EPFL, Lausanne (CH), 1983-1984.

[56] G. G. Stokes, *On the theory of oscillatory waves*, Trans. Camb. Phil. Soc. **8** (1847), 441–455.

[57] ———, *Considerations relative to the greatest height of oscillatory irrotational waves which can be propagated without change of form*, Mathematical and Physical Papers, vol. I, Cambridge, 1880, pp. 225–228.

[58] A. E. Taylor, *Introduction to Functional Analysis*, John Wiley, New York, 1958.

[59] E. C. Titchmarsh, *Theory of Functions*, 2nd. ed., Oxford University Press, 1939.

[60] J. F. Toland, *Stokes waves*, Topological Methods in Nonlinear Analysis **7 & 8** (1996 & 1997), 1–48 & 412–414.

[61] ———, *Calculus and the Implicit Function Theorem*, lecture notes, University of Bath (U.K.), 1996.

[62] ———, *Stokes waves in Hardy spaces and as distributions*, J. Mat. Pure et Appl. **79 (9)** (2000), 901–917.

[63] _____, *A pseudo-differential equation for Stokes waves*, Arch. Rational Mech. Anal. **162** (2002), 179–189.

[64] R. E. L. Turner, *Transversality and cone maps*, Arch. Rational Mech. Anal. **58** (1975), 151–179.

[65] B. L. Van der Waerden, *Modern Algebra*, Ungar Publishing Company, New York, 1949.

[66] H. Whitney, *Complex Analytic Varieties*, Addison-Wesley, Reading, Mass. USA, 1972.

[67] J. Wloka, *Partial Differential Equations*, Cambridge University Press, Cambridge, 1987.

[68] E. Zeidler, *Nonlinear Functional Analysis and its Applications I*, Springer-Verlag, New York, 1985.

[69] A. Zygmund, *Trigonometric Series I & II*, corrected reprint (1968) of 2nd. ed., Cambridge University Press, Cambridge, 1959.

Index

9 780691 112985